Theory of Special Relativity for Beginners

Jen Chyi Liao

Contact: liao_jen@hotmail.c

Publisher: Amazon Kindle Direct Publishing

Contents

Chapter1: Introduction

Albert Einstein published two postulates in 1905 with the aim of resolving the issue of the propagation of electromagnetic radiation. A full-blown theory, known as the theory of special relativity today, based on the two postulates was then developed. The theory can be regarded as an extension of Newtonian physics (also known as Newtonian mechanics). Whereas Newtonian physics fails at speeds close to c, the speed of light, the theory of special relativity (the Relativity for short from now on) covers speeds from zero to c.

The Relativity is one of the greatest achievements of human intellect, and is often regarded as a subject difficult to comprehend. Since the Relativity is an extension of Newtonian physics, every aspects of the Relativity are in parallel to those of Newtonian physics. As Newtonian physics is comprehensible to most people, so should the Relativity be. The Relativity may be difficult to comprehend simply because it is not presented in an easy to understand format.

As this book is intended for beginners, some topics that were covered in some introductory books are deliberately excluded. Those topics include geometric representation of space-time, Minkowski's four-dimensional space-time, etc.

The mathematics in this book involves only basic calculus. However, those who are not comfortable with it or are not interested in mathematical details can skip or glance through without missing the essential part of the theory.

1.1 Newtonian Physics and the Relativity

In physics, there is an important principle called the "principle of special relativity" which requires that all laws of physics have the same mathematical forms in all inertial systems. Newtonian physics and the Relativity obey the principle of special relativity, and are both a theory of special relativity. However, today the term "theory of special relativity" usually refers to the one proposed by Einstein.

Although Newtonian physics obeys the principle of special relativity, the one it obeys is a limited version in that the theory for electromagnetic radiation is not included. Newtonian physics and the Relativity are also different in the underlying concepts that they are built upon. Newtonian physics is built upon the concept of absolute space and absolute time. The Relativity is built upon the postulate that the speed of light in the vacuum has the same value c in all inertial systems. The postulate of the constancy of light speed is in contradiction with the concept of absolute space and absolute time.

1.2 Reference Frame and Inertial System

In physics, we study physical phenomena by observing (measuring) them. A physical phenomenon consists of a series of physical events. An event happens at a point in space and at an instant of time. The space location and time of occurrence of an event is called the space-time data of that event. Observing a physical phenomenon involves taking the space-time data as well as physical quantities such as mass and electric charge of all objects involved in each event. Thus, in order to observe a physical phenomenon we need to take and collect the space-time data. How are these data taken and collected?

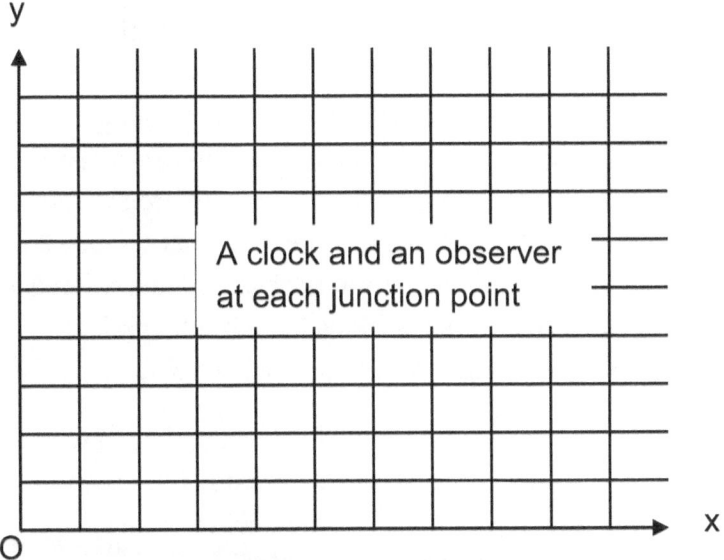

Fig. 1.1 Reference frame

Imagine that there exists a three-dimensional, lattice-like structure with its two-dimensional view as shown in Fig. 1.1. The distance of each junction point from a reference point (the origin) has been marked by a measurement using standard meters, and there is a clock and an observer at each junction point. All clocks have been synchronized and calibrated against one another before being distributed and placed at the junction points. Thus, the time that an object arrives at any space point can be recorded by an observer. In other words, with such a structure, the space-time data of any object can be obtained.

Although such a structure cannot be physically implemented as it would obstruct motion of objects, there are means for taking the space-time data as if such a structure were in place. Thus, we can regard such an imaginary structure as a reference frame for observing physical phenomena. When a physical phenomenon is observed in a place, such as a laboratory on the ground or a moving train, it is assumed that there is a reference frame attached to that place. We can simply say that the place in which the physical phenomenon is observed is a reference frame. A reference frame is like a huge mathematical coordinate system with observers attached to it.

An inertial system is a reference frame at rest or moves rectilinearly with a constant speed. We don't know the absolute motion of any object in the universe; therefore, we cannot determine if an object is moving rectilinearly with a constant speed. Nonetheless, physicists take a laboratory fixed on the ground of the earth as an inertial system although the earth is orbiting about the Sun and is also rotating about an axis of its own. Thus, a laboratory on the ground is an inertial system, and anything moving at a constant velocity with respect to the laboratory is also an inertial system, but things that move with varying speed or direction are not.

Although inertial systems can move only with uniform velocity with respect to one another, an object under study may have any kind of velocity. In other word, the speed and direction of the object can change.

An event can be observed in different inertial systems simultaneously. The space-time data of an event in an inertial system are different from those in other inertial systems.

If an event is observed in two inertial systems denoted as S and S', respectively; a set of space-time data (x, y, z, t) is obtained in S and another set of space-time data (x', y', z', t') is obtained in S'. In Newtonian physics, the two sets of data are related by a set of equations called the Galilean transformation equations, or simply the Galilean transformations.

1.3 Absolute Space and Absolute Time

Most of us have the notion of an empty space which is at rest and in which everything of the universe move. Notion of such an empty space is the concept of absolute space.

Since the empty space is at rest, it is an inertial system. Anything moving at a constant velocity relative to the empty space is also an inertial system. The empty space is unique or superior in that it can be used to determine the absolute speeds of other reference frames. All inertial systems other than the empty space are equivalent as none of them can be considered as superior than others.

The absolute time is just the perception of time in our daily experience. We read time from clocks, and there are different kinds of clocks such as mechanic, electronic, etc. All clocks can be calibrated and synchronized at the same place using a standard clock and then distributed to different locations. It is assumed that operation and accuracy of clocks are not affected by transportations of any kind. Clocks and time of this kind are thus universal. Events at different locations are said to be simultaneous if the clocks at those locations have the same reading when the events occur.

1.4 Galilean Transformations

The Galilean transformations are derived based on the concept of absolute space and absolute time.

Assume an inertial system S' moving in the x-direction at a speed v relative to another inertial system S. This is a legitimate assumption without lose of generality as we can always rotate S and S' so that S' moves in the x-direction. We also assume that the origin and all spatial axes of S' coincide with those of S at $t = t' = 0$ as shown in Fig. 1.2.

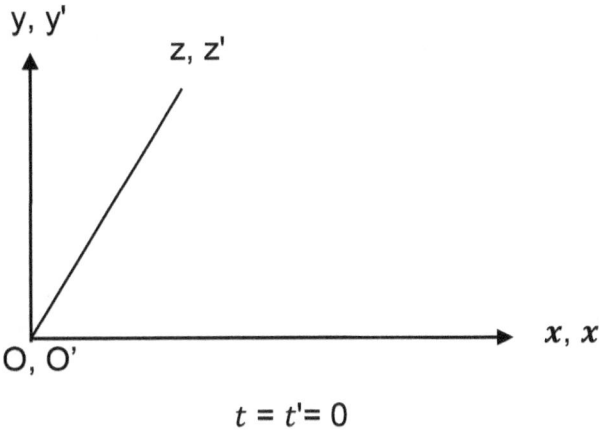

$$t = t' = 0$$

Fig. 1.2 S and S' at $t = t' = 0$

At $t = t' = 0$, the space-time data of any event is the same in S and S'; but as time elapses, this will change because S' is moving with respect to S. The situation is illustrated in Fig. 1.3.

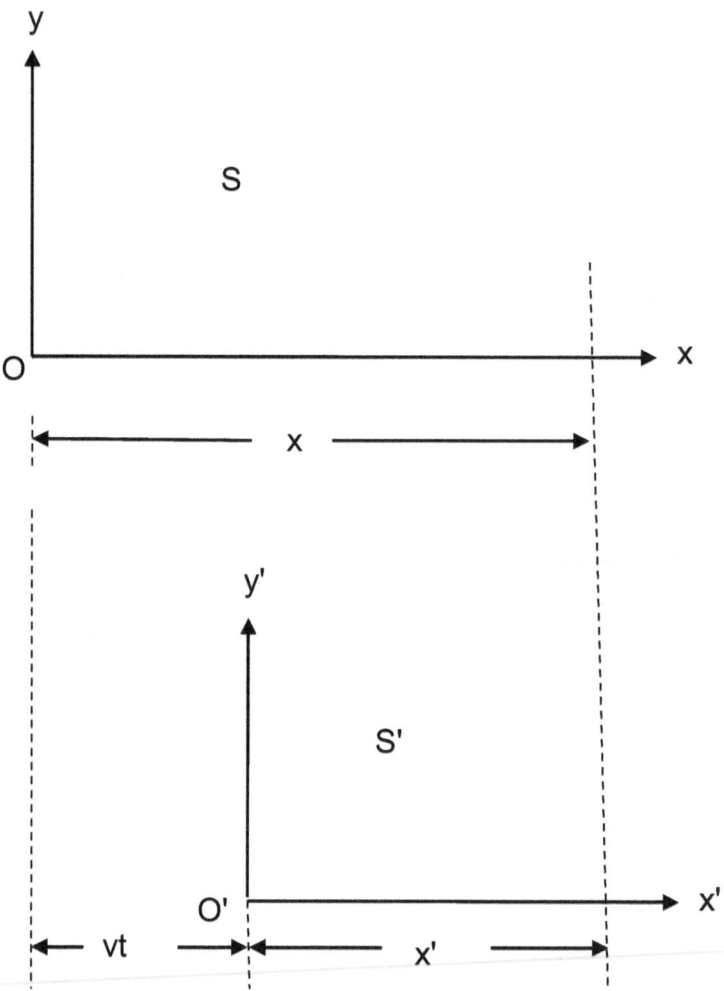

Fig. 1.3 S and S' at t = t' > 0

When $t = t' > 0$, it can be seen that

$$x = x' + vt$$

Since $t = t'$, the above equation can be written as

$$x = x' + vt' \quad (1.4.1)$$

Since S' moves in the x-direction only,

$$y = y' \quad (1.4.2)$$

$$z = z' \quad (1.4.3)$$

Because of the universal nature of time,

$$t = t' \quad (1.4.4)$$

Equations (1.4.1) - (1.4.4) are for transformation from S' to S. Equations for transformation from S to S' can be obtained in a similar way by considering that S' is at rest and S is moving at a speed v in the -x-direction, and the result is

$$x' = x - vt \quad (1.4.5)$$

$$y' = y \quad (1.4.6)$$

$$z' = z \quad (1.4.7)$$

$$t' = t \quad (1.4.8)$$

Since the velocity of an object is the time derivative of the position of the object, transformation equations for velocity can be obtained by taking the time derivative of Equations (1.4.1) - (1.4.3).

Differentiating Equation (1.4.1) with respect to t we obtain

$$\frac{dx}{dt} = \frac{dx'}{dt} + v\frac{dt'}{dt}$$

Since t = t', the above equation can be written as

$$\frac{dx}{dt} = \frac{dx'}{dt'} + v \quad (1.4.9)$$

Using the same approach we can obtain

$$\frac{dy}{dt} = \frac{dy'}{dt'} \quad (1.4.10)$$

and

$$\frac{dz}{dt} = \frac{dz'}{dt'} \quad (1.4.11)$$

Equations (1.4.9) - (1.4.11) can also be written as

$$u_x = u'_x + v \quad (1.4.12)$$

$$u_y = u'_y \quad (1.4.13)$$

$$u_z = u'_z \quad (1.4.14)$$

In the above equations, u_x is the x-component of the velocity in S, and u'_x is the x-component of the velocity in S', and so on.

Since the acceleration of an object is the time derivative of the velocity of the object, transformation equations for acceleration can be obtained using the approach we just used for velocity transformations. The result is

$$a_x = a'_x \quad (1.4.15)$$

$$a_y = a'_y \quad (1.4.16)$$

$$a_z = a'_z \quad (1.4.17)$$

Since the acceleration is the same in all inertial systems, the second law of Newtonian physics

$$F = ma$$

where F is the force and m is the mass of an object, is valid in all inertial systems provided that the mass of an object is a constant quantity.

1.5 Time Interval under Galilean Transformations

In physics, we are often more interested in the space and time intervals between two events than the space-time data of an individual event.

In S, the time interval T between two events that occur at t_1 and t_2 is

$$T = t_2 - t_1$$

In S', the two events occur at t'_1 and t'_2 and the time interval T' is

$$T' = t'_2 - t'_2$$

According to the Galilean transformations

$$t_1 = t'_1$$

$$t_2 = t'_2$$

Thus,

$$T = T'$$

The time intervals are the same in both systems. As this is true for all inertial systems, the time interval between two events is invariant in all inertial systems.

1.6 Space Interval under Galilean Transformations

If we want to measure the length of an object or the space interval (distance) between two space points, how shall we proceed? If the object is rest, we can get the space data of the two ends at any time and then calculate from the data. But if the object is moving, then we have to get the space data of the two ends at the same time; otherwise the result would be erroneous.

A meter stick is placed along the x-axis and is at rest in S' which is moving at a speed v in the x-direction with respect to S. The two ends of the meter stick are at x'_1 and x'_2 in S', and are at x_1 and x_2 in S. According to the Galilean transformations,

$$x'_1 = x_1 - vt_1$$

$$x'_2 = x_2 - vt_2$$

Thus,

$$x'_2 - x'_1 = x_2 - x_1 - v(t_2 - t_1)$$

Since the space data are taken at the same time, i.e. $t_1 = t_2$,

$$x'_2 - x'_1 = x_2 - x_1$$

As $(x'_2 - x'_1)$ is the length of the meter as well as the space interval between two space points in S', and $(x_2 - x_1)$ is the length of the meter as well as the space interval between two space points in S; we can conclude that space interval is invariant in all inertial systems.

1.7 The Law of Velocity Addition

If u and v are in the same direction but not limited to the x-direction, then Equation (1.4.12) becomes

$$u = u' + v \quad (1.7.1)$$

Equation (1.7.1) is the law of velocity addition of Newtonian physics. For example, if a passenger on a train, which moves at a speed of 60 miles per hour, runs with a speed of 5 miles per hour toward the head of the train, the speed of the passenger observed by an observer on the ground is then 65 miles per hour.

1.8 Time Order and Simultaneity of Events

Because of the universal nature of time in Newtonian physics, time orders of events are the same in all inertial systems. Events that are simultaneous in an inertial system are also simultaneous in all other inertial systems.

1.9 Form of the Law of Physics

In Newtonian physics, space and time intervals are invariants, and the mass of an object is assumed to be a constant quantity. Thus, the three basic quantities in physics, i.e. length, time, and mass, are all invariants; therefore, the law of physics has the same form in all inertial systems.

Chapter 2: Postulates of the Relativity

Although the Relativity is an extension of Newtonian physics; when Einstein presented the two underlying postulates of the Relativity, his aim was to resolve the issue of the propagation of electromagnetic radiation (wave). In this chapter, we are going to give an account of the issue and some efforts that had been done to resolve it at the time Einstein presented his postulates.

2.1 Maxwell's Theory of Electromagnetism and Ether

In 1864, James Clark Maxwell combined the theories of electricity and magnetism into a comprehensive theory known as Maxwell's theory or Maxwell's equations today.

Maxwell's theory reveals that light is a form of electromagnetic radiation. The theory also predicts that light propagates in the vacuum with a speed c and

$$c = \frac{1}{\sqrt{\epsilon_0 \mu_0}} \qquad (2.1.1)$$

where ε_0 is the permittivity of free space and μ_0 is the permeability of free space, and the value of c is approximately 3×10^8 m/s.

There are two problems with Maxwell's theory. First, it does not specify the medium through which the electromagnetic radiation propagates. Second, it does not specify in which inertial system the value of c is obtained. According to the law of velocity addition of Newtonian physics, the speed of light is different in different inertial systems. But Maxwell's theory implies that the speed of light is a constant and is thus the same in all inertial systems.

At the time Maxwell presented his theory, physicists firmly believed that electromagnetic radiation need a medium for propagation. A substance called "ether" was proposed as the medium. The ether is at rest and fills the space of the universe.

If the ether is indeed the medium of propagation of electromagnetic radiation, the speed of light is of the value c only when observed in an inertial system at rest with respect to the ether. In an inertial system moving with a speed v relative to the ether, the observed speed of light should be either $(c + v)$ or $(c - v)$. The ether is an inertial system that is unique in that it is the only one in which the speed of light is c.

Physicists wanted to verify the existence of ether by experiments. The essential idea was to see if light speeds measured in different inertial systems are different and to see if there is a unique inertial system in which the light speed is c. This requires inertial systems that move with speeds comparable to the light speed. The maximum speed attainable on the earth is the speed of the earth around the sun, 3×10^4 m/s.

Assuming the earth is moving in the ether, the speed v of an inertial system attached to the earth is then 3×10^4 m/s. The deviation of the speed of light measured in this inertial system is only 0.01% of the speed of light. At the time the experiment was proposed, there was no equipment that can measure such a small fractional change accurately.

The experiment was finally performed in 1887 by A. A. Michelson and E. W. Morley using a precision optical interferometer developed by A. A. Michelson a few years earlier.

2.2 Michelson-Morley Experiment

The basic principle of Michelson-Morley experiment is using a setup shown in Fig. 2.1. The setup consists of four mirrors, M1 - M4, fixed in a laboratory on the ground of the earth. The distance between M1 and M2 is the same as that of M3 and M4, and the centerline of M1 and M2 is perpendicular to that of M3 and M4. A light beam is emitted from a source at M1, reflected by M2 and then returns to M1. The time it takes is t_x. A light beam is emitted from a source at M3, reflected by M4 and then returns to M3. The time it takes is t_y. If the ether hypothesis is correct, t_x and t_y might be different.

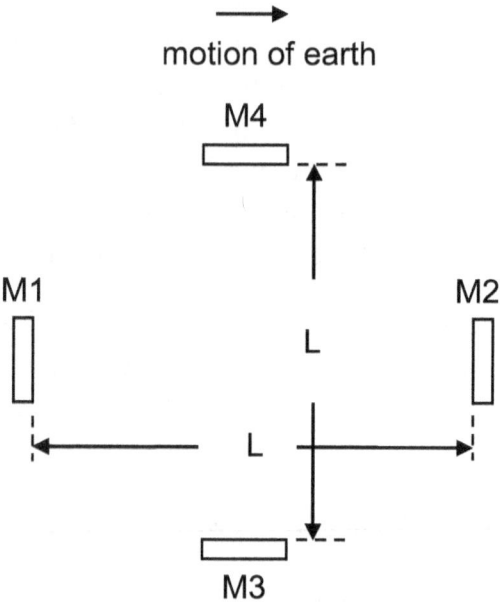

Fig. 2.1 Setup of Michelson-Morley experiment

We first derive the expression for t_y in the ether system. Since both M3 and M4 are moving to the right at a speed v, only light beam emitted at M3 with an angle of θ as shown in Fig. 2.2 can reach M4. The same is true for the light beam reflected from M4 and then returns to M3.

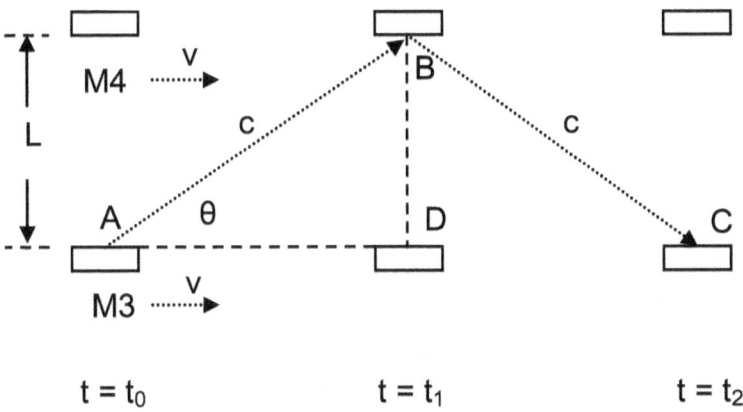

Fig. 2.2 Derivation of t_y in the ether system

The light beam leaves M3 at time t_0, arrives at M4 at t_1, and returns to M3 at t_2. It can be seen from Fig. 2.2 that

$$t_y = t_2 - t_0 = 2(t_1 - t_0) \quad (2.2.1)$$

$$AD = v(t_1 - t_0) = \frac{1}{2}vt_y \quad (2.2.2)$$

$$AB = c(t_1 - t_0) = \frac{1}{2}ct_y \quad (2.2.3)$$

$$BD = L \quad (2.2.4)$$

Since the triangle ABD is a right triangle, by the Pythagorean Theorem we obtain

$$(AB)^2 = (AD)^2 + (BD)^2 \quad (2.2.5)$$

Theory of Special Relativity for Beginners

From Equations (2.2.1) - (2.2.5) we obtain

$$t_y = \frac{2L}{c} \frac{1}{\sqrt{1 - \frac{v^2}{c^2}}} \qquad (2.2.6)$$

We next derive the expression for t_x in the ether system. Both M1 and M2 are moving to the right at a speed v as shown in Fig. 2.3. The light beam leaves M1 at t_0 and arrives at M2 at t_1. During this time period M2 moves a distance l_1 which is equal to $v(t_1 - t_0)$, and the light beam travels a distance of $(L + l_1)$. The light beam is reflected by M2 at time t_1 and returns to M1 at time t_2. During this time period, M1 moves a distance l_2 which is equal to $v(t_2 - t_1)$, and the light beam travels a distance of $(L - l_2)$.

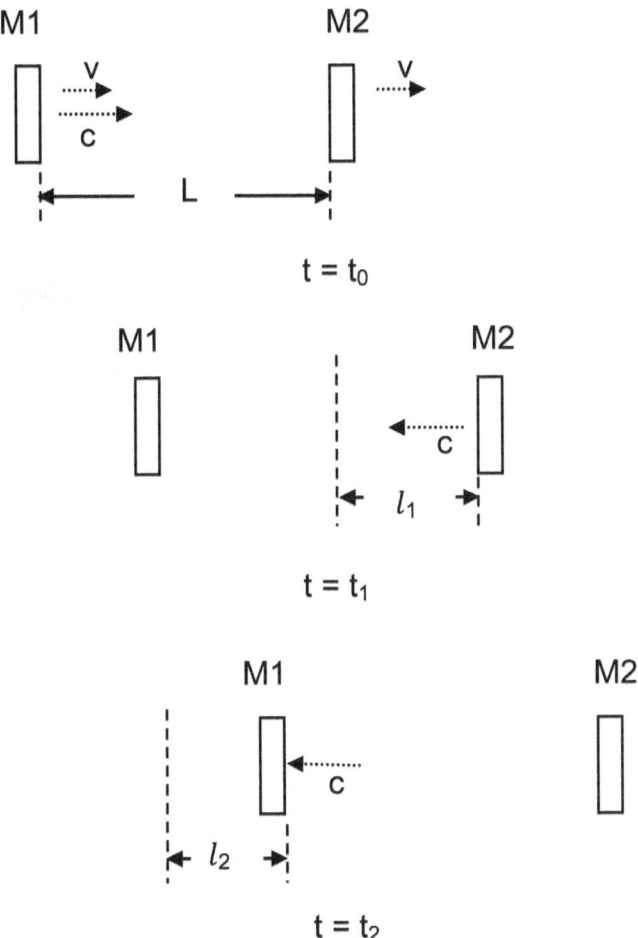

Fig. 2.3 Derivation of t_x in the ether system

We can also obtain the following relationships from Fig. 2.3

$$t_x = t_2 - t_0 = (t_2 - t_1) + (t_1 - t_0) \quad (2.2.7)$$

$$c(t_1 - t_0) = L + l_1 = L + v(t_1 - t_0) \quad (2.2.8)$$

$$c(t_2 - t_1) = L - l_2 = L - v(t_2 - t_1) \quad (2.2.9)$$

Combining Equations (2.2.7) – (2.2.9) we obtain

$$t_x = \frac{2L}{c} \frac{1}{1 - \frac{v^2}{c^2}} \quad (2.2.10)$$

It is clear that t_x and t_y have different expressions and will consequently have different values. Thus, after being reflected and returned to the point of emission the two light beams will have different phases. If the two beams are combined, the resultant light beam will show certain interference pattern on Michelson's interferometer.

Michelson and Morley performed the experiment at different times of a day and in all seasons of a year. In other words, the experiment was observed in different inertial systems corresponding to different states of motion of the earth. No predicted interference pattern was observed in any of the experiments.

The derivation of the expressions for t_x and t_y is based on the assumption that the ether exists and that the mirrors are moving with respect to the ether. The result of Michelson-Morley experiment did not agree with the prediction based on the above assumptions. Thus, Michelson-Morley experiment failed to confirm the existence of ether.

2.3 Efforts to Save the Ether Hypothesis

Despite of the result of Michelson-Morley experiment, physicists were reluctant to give up the idea of ether. Several theories were proposed in order to save the ether hypothesis. Some of them will be described in the following sections.

2.3.1 Lorentz-Fitzgerald Contraction Hypothesis

George Francis Fitzgerald and Hendrick Lorentz independently proposed the so-called "contraction hypothesis". According to this hypothesis, when a body moves at a speed v relative to the ether, its length in the direction of motion is contracted as shown in Fig. 2.4.

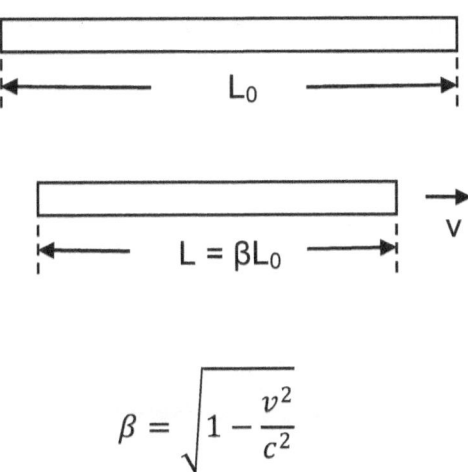

$$\beta = \sqrt{1 - \frac{v^2}{c^2}}$$

Fig. 2.4 Length contraction due to motion

In the Michelson-Morley experiment, the mirrors are mounted on an apparatus which is moving with respect to the ether. Whereas the distance (length) between M1 and M2 is in the direction of motion, that between M3 and M4 is not. Thus, the length L in Equation (2.2.10) must be contracted, and the expression for t_x is the same as that for t_y.

Although the Lorentz-Fitzgerald contraction hypothesis can account for the result of Michelson-Morley experiment, it failed to do so for other similar experiments. Moreover, there is no experimental evidence of length contraction at all.

2.3.2 Ether-Drag Hypothesis

This hypothesis assumes that the ether surrounding a body, such as the earth, is dragged along by the body as shown in Fig. 2.5. Thus, the value of v in Equations (2.2.6) and (2.2.10) becomes zero, and the expression for t_x is the same as that for t_y.

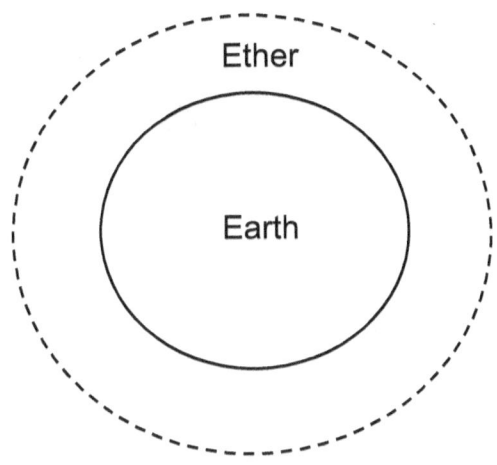

Fig. 2.5 Ether moves with the earth

Although the ether-drag hypothesis can account for the result of Michelson-Morley experiment, it contradicts a phenomenon called "stellar aberration". When one is to observe a star directly overhead using a telescope as shown in the left side of Fig. 2.6; because of the motion of the earth, he or she has to tilt the telescope as shown in the right side of Fig. 2.6, otherwise the light from the star would hit the wall of the telescope and cannot reach the eye. The observed position of the star is thus the apparent one instead of the real one. If the ether surrounding the earth is dragged along by the earth; the light from the star would be carried by the ether and would be at rest with respect to the earth, and there would be no stellar aberration. Thus, the ether-drag hypothesis is untenable.

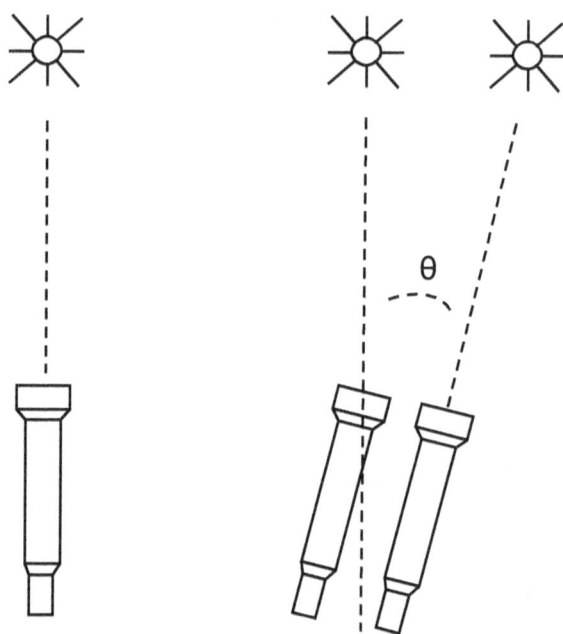

Fig. 2.6 Stellar aberration

2.3.3 Emission Theories

Some physicists proposed hypotheses collectively called "emission theories". The essential idea is that the speed of light depends on the state of motion of the source. The speed of light is c only when the source is at rest with respect to the ether. The speed of light from a source that moves at a speed v with respect to the ether is c + v in the ether system. Light emitted from a source is, in a sense, carrying the inertia of the source.

Although the emission theories can account for the result of Michelson-Morley experiment, all experiments showed that the speed of light is c in all inertial systems. The emission theories are thus untenable.

2.4 Postulates of the Relativity

At the beginning of the twentieth century, physicists have failed to identify a unique inertial system in which the speed of light is c. In addition, all experiments showed that the speed of light is c in all inertial systems. In 1905, Einstein presented two postulates which can be stated as the following:

Poatulate1: the law of physics is the same in all inertial systems.

Postulate 2: the speed of light is the same in all inertial systems.

The two postulates are phrased somewhat differently in different books, but the meaning is the same.

The first postulate is just a reiteration of the principle of special relativity we have discussed. The second postulate suggests that we accept the constancy of light speed as a fact of nature. We need not to invent something like the ether for the propagation of light. Consequently, we need not to verify the existence of ether and need not be concerned about the result of Michelson-Morley experiment.

The two postulates reveal nothing more than existing knowledge at that time. The remarkable thing is that Einstein realized that the Galilean transformations contradict the postulates and a new set of transformation equations is required. Einstein also found that the forms of some laws of physics need to be modified, and a full-blown theory was then developed.

2.5 Einstein's Concept of Time and Space

Einstein's concept of time and space is very different from that of Newtonian physics. From his viewpoint, time and space, like other properties of physics, are quantities obtained by measurement. Measurement involves signal transmission which can be delayed depending on the distance between the event and the observer. This can be illustrated using a simple example shown in Fig. 2.7.

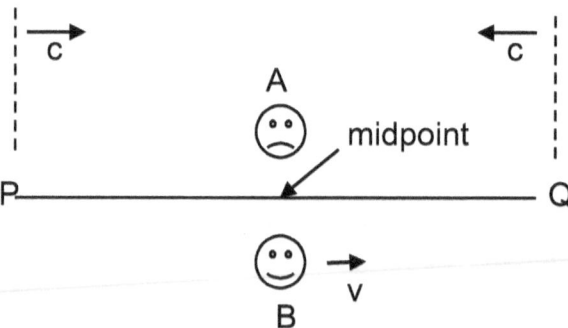

Fig. 2.7 Relation between event and observers

In Fig. 2.7, P and Q are two space locations on a straight road. Observer A stands at the midpoint M between P and Q, and Observer B is running to the right along the road. At the instant that Observer B passes by Observer A, explosions occur simultaneously at P and Q and lights propagate toward M from both locations. Observer A stands at the midpoint; he or she will receive the signals at the same time, and will judge that the explosions occur simultaneously. Observer B is running toward Q; he or she will receive the signal from Q first, and will judge that the explosion at Q occurs prior to that of P. In other words, the time perceived by the two observers is different.

Physics is a science of measurement and the laws of physics are for predicting quantities that are measurable. The universal time (absolute time) in Newtonian physics is not obtained by measurement. Einstein decided to abandon the concept of absolute time and space, and develop a new set of laws based on the two postulates he proposed.

Chapter 3: Lorentz Transformations and Kinematics of the Relativity

We have learned that in Newtonian physics the space-time data of two inertial systems are related by a set of equations called the Galilean transformations. In this chapter, we are going to derive a new set of transformation equations based on Einstein's two postulates and compare it with the Galilean transformations. We will also show how a new law of motion, i.e. kinematics of the Relativity, is resulted.

3.1 Lorentz Transformations

The new set of transformation equations can be derived in a similar way as the Galilean transformations. We first made the same assumptions for S and S' as in the case of the Galilean transformations and set the origin of S' to coincide with the origin of S at t = t' = 0. As time elapses, the relation between S and S' is shown in Fig. 3.1.

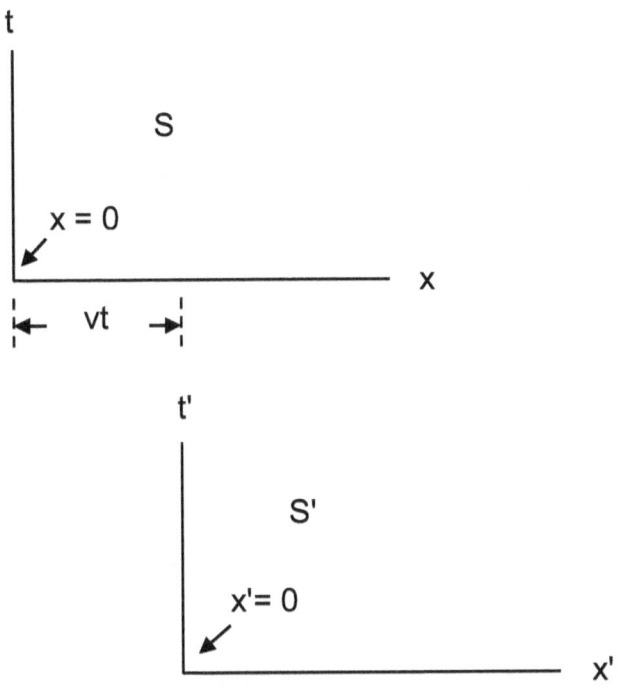

Fig. 3.1 Relation between S and S'

An event in S must correspond to a single event in S', and vice versa. To ensure the 1-to-1 correspondence, all equations must be linear, i.e. they must involve only the first power in the variables. In addition, t' must be independent of y and z, and t must be independent of y' and z'.

According to the first postulate of the Relativity, the laws of physics must be the same in all inertial systems. Thus, a subject moves in a direction parallel to the x-axis in S must also move in a direction parallel to the x'-axis in S'. This requires that

$$y' = y$$

$$z' = z$$

Thus, the equations for transforming from S to S' should be as the following:

$$x' = ax + et \quad (3.1.1)$$

$$y' = y \quad (3.1.2)$$

$$z' = z \quad (3.1.3)$$

$$t' = bx + dt \quad (3.1.4)$$

In the above equations, a, b, d, and e are coefficients to be determined.

Since S' is moving in the x-direction at a speed v, the location x' = 0 in S' corresponds to the location $x = vt$ in S as shown in Fig. 3.1. Combining this relation with Equation (3.1.1) we obtain

$$0 = avt + et$$

$$e = -av \quad (3.1.5)$$

Inserting Equation (3.1.5) into Equation (3.1.1) we obtain

$$x' = a(x - vt) \quad (3.1.6)$$

Suppose that at t = 0 a light source at the origin emits a flash of light. The light then spread out in all directions as shown in Fig. 3.2. The state of motion of the light source is not a concern here as we are only interested in the light that is spreading out from the origin.

Since the origins of S and S' coincide at t = t' = 0; from the viewpoint of observers in S, the light spread out with the origin O as the center; whereas from the viewpoint of observers in S', the light spread out with the origin O' as the center. The above is not only true at t = t' = 0 but also true as time elapses.

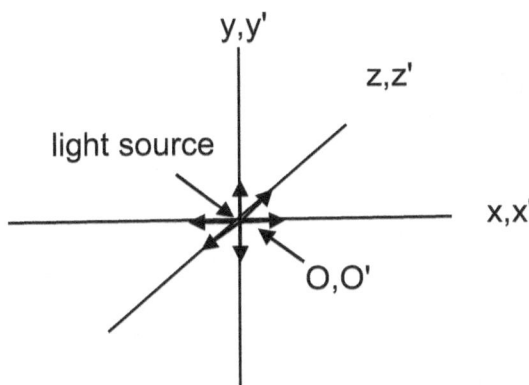

Fig. 3.2 Light emitted from the origin at t = t' = 0

When the light spread out, the distribution of the wave-front, i.e. the front-most points of the wave, is spherical as shown in Fig. 3.3. According to the first postulate of the

Relativity, this holds true in both S and S'. The expression for the wave-front in S is

$$x^2 + y^2 + z^2 = c^2t^2 \quad (3.1.7)$$

where c is the speed of light. Similarly, the expression for the wave-front in S' is

$$x'^2 + y'^2 + z'^2 = c^2t'^2 \quad (3.1.8)$$

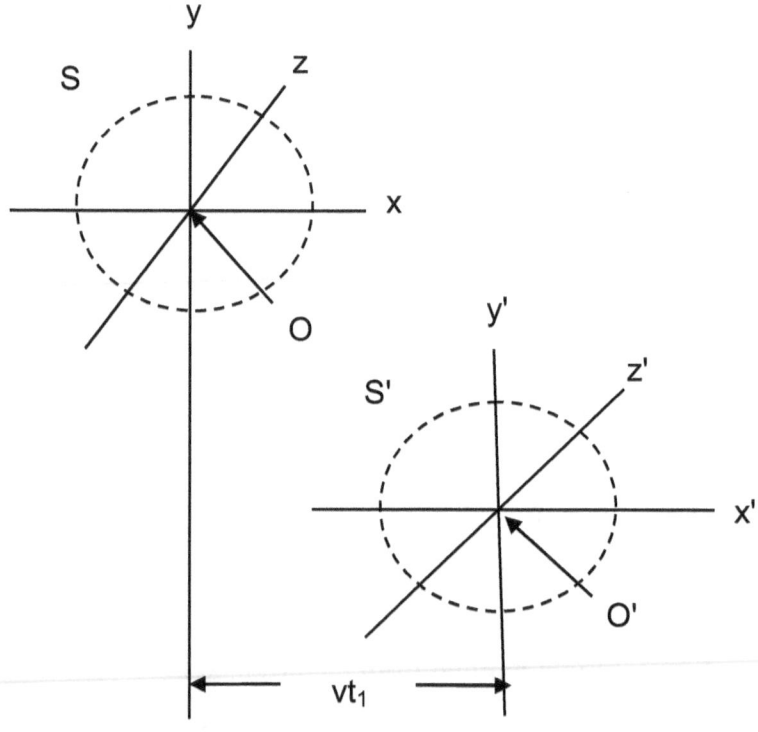

Fig. 3.3 Wave-front distribution at $t = t_1$

Inserting Equations (3.1.2), (3.1.3), (3.1.4) and (3.1.6) into Equation (3.1.8) we obtain

$$(a^2 - c^2 b^2)x^2 + y^2 + z^2 - 2(va^2 + c^2 bd)xt$$
$$= (c^2 d^2 - v^2 a^2)t^2 \quad (3.1.9)$$

By comparing Equation (3.1.7) with Equation (3.1.9) we obtain the following relations

$$a^2 - c^2 b^2 = 1 \quad (3.1.10)$$

$$va^2 + c^2 bd = 0 \quad (3.1.11)$$

$$c^2 d^2 - v^2 a^2 = 1 \quad (3.1.12)$$

Coefficient a, b and d can be determined from the above equations and are

$$a = \frac{1}{\sqrt{1 - \frac{v^2}{c^2}}} \quad (3.1.13)$$

$$b = \frac{-(\frac{v}{c^2})}{\sqrt{1 - \frac{v^2}{c^2}}} \quad (3.1.14)$$

$$d = \frac{1}{\sqrt{1 - \frac{v^2}{c^2}}} \quad (3.1.15)$$

The set of transformations equations are

$$x' = \frac{x - vt}{\sqrt{1 - \frac{v^2}{c^2}}} \quad (3.1.16)$$

$$y' = y \quad (3.1.17)$$

$$z' = z \quad (3.1.18)$$

$$t' = \frac{t - \left(\frac{v}{c^2}\right)x}{\sqrt{1 - \frac{v^2}{c^2}}} \quad (3.1.19)$$

Since these equations have the same form as those presented by H. A. Lorentz before Einstein published his theory of special relativity, they are called the Lorentz transformation equations or simply the Lorentz transformations.

Equations (3.1.16) and (3.1.19) can be written as

$$x' = \gamma(x - vt) \quad (3.1.20)$$

$$t' = \gamma\left(t - \frac{vx}{c^2}\right) \quad (3.1.21)$$

where

$$\gamma = \frac{1}{\sqrt{1 - \beta^2}} \quad (3.1.22)$$

and

$$\beta = \frac{v}{c} \quad (3.1.23)$$

β is called the speed parameter and γ the Lorentz factor.

Equations (3.1.16) - (3.1.19) are transformations from S to S'. Transformation equations from S' to S can be obtained in the same way as we did for the Galilean transformations, i.e. interchanging primed and unprimed coordinates and replacing v by -v. The result is

$$x = \gamma(x' + vt') \quad (3.1.24)$$

$$y = y' \quad (3.1.25)$$

$$z = z' \quad (3.1.26)$$

$$t = \gamma\left(t' + \frac{vx'}{c^2}\right) \quad (3.1.27)$$

The space and time intervals in S between an event at (t_1, x_1, y_1, z_1) and another one at (t_2, x_2, y_2, z_2) are

$$\Delta t = t_2 - t_1$$

$$\Delta x = x_2 - x_1$$

$$\Delta y = y_2 - y_1$$

$$\Delta z = z_2 - z_1$$

The space and time intervals in S' between the same two events are

$$\Delta t' = t'_2 - t'_1$$

$$\Delta x' = x'_2 - x'_1$$

$$\Delta y' = y'_2 - y'_1$$

$$\Delta z' = z'_2 - z'_1$$

Using the Lorentz transformations for space-time data we obtain the following transformation equations for space and time intervals

$$\Delta x' = \gamma(\Delta x - v\Delta t) \quad (3.1.28)$$

$$\Delta y' = \Delta y \quad (3.1.29)$$

$$\Delta z' = \Delta z \quad (3.1.30)$$

$$\Delta t' = \gamma(\Delta t - \frac{v\Delta x}{c^2}) \quad (3.1.31)$$

$$\Delta x = \gamma(\Delta x' + v\Delta t') \quad (3.1.32)$$

$$\Delta y = \Delta y' \quad (3.1.33)$$

$$\Delta z = \Delta z' \quad (3.1.34)$$

$$\Delta t = \gamma(\Delta t' + \frac{v\Delta x'}{c^2}) \quad (3.1.35)$$

3.2 Comparison with Galilean Transformations

Equations of Lorentz and Galilean transformations are listed for a comparison in Table 3.1.

Table 3.1 Lorentz transformations versus Galilean transformations

Galilean transformations	Lorentz transformations
$x' = x - vt$	$x' = \gamma(x - vt)$
$y' = y$	$y' = y$
$z' = z$	$z' = z$
$t' = t$	$t' = \gamma\left(t - \frac{vx}{c^2}\right)$
$\Delta x' = \Delta x$	$\Delta x' = \gamma(\Delta x - v\Delta t)$
$\Delta t' = \Delta t$	$\Delta t' = \gamma(\Delta t - \frac{v\Delta x}{c^2})$

In the Galilean transformations, time coordinate and space coordinate are independent from each other. In contrast, in the Lorentz transformations, time coordinate and space coordinate are interrelated, and are both related to the relative speed v of the inertial systems.

In the Galilean transformations, time interval and space interval are not related to each other and are independent of the inertial system. In contrast, in the Lorentz transformations, time interval and space interval are interrelated, and are both related to the relative speed v of the inertial systems.

When v is much smaller then c,

$$\beta = \frac{v}{c} \approx 0$$

and

$$\gamma \approx 1$$

Under this condition, the two transformations are almost the same. The Galilean transformations can be regarded as an approximation to the Lorentz transformations at low speed.

At what speed should we be concerned about the validity of the Galilean transformations? Let us take a look of speeds of various objects shown in Table 3.2.

Table 3.2 Object speed and corresponding value of γ

Speed	β	γ	Object
110 km/h	1.033×10^{-7}	≈ 1	Automobile on free way
400 km/h	3.7×10^{-7}	≈ 1	Sports car
600 km/h	4.63×10^{-7}	≈ 1	Airplane
1,000 km/h	9.27×10^{-7}	≈ 1	Supersonic airplane
11,300 km/h	1.046×10^{-5}	≈ 1	Man-made satellite
0.3×10^8 m/s (0.1c)	0.1	1.005	High-energy particle
1.5×10^8 m/s (0.5c)	0.5	1.155	High-energy particle
2.97×10^8 m/s (0.99c)	0.99	7.089	High-energy particle

It can be seen that the speed of object in our daily experience is much lower than c. Only in the region of high-energy particles we need to be concerned about the validity of the Galilean transformations or equivalently Newtonian physics.

Now let us make a numerical comparison between the Galilean and Lorentz transformations using an example in which a "c-speed-train" can reach a speed close to c. The c-speed-train represents a group of high-energy particles moving in a particle accelerator, and the train station represents the control center of the lab.

Theory of Special Relativity for Beginners

A c-speed-train moves at a speed of 0.5c in the x-direction through the train station. Let S' be the inertial system at rest with respect to the train and S be the inertial system at rest with respect to the platform of the train station. An event occurs at x = 10,000 m and t = 30 x 10^{-6} s in S. We want to find out the corresponding space-time data in S'.

It should be pointed out that in this kind of analysis the origins of S and S' are always coincident at t = t' = 0.

The results from the Galilean transformations are

$$x' = x - vt = 10,000 - 1.5 \times 10^8 \times 30 \times 10^{-6}$$
$$= 10,000 - 4,500 = 5,500$$

$$t' = t = 30 \times 10^{-6}$$

The results from the Lorentz transformations are

$$\beta = \frac{v}{c} = 0.5$$

$$\gamma = \frac{1}{\sqrt{1 - (0.5)^2}} = 1.1547$$

$$x' = \gamma(x - vt) = 1.1547 \times 5,500 = 6,350.9$$

$$t' = \gamma\left(t - \frac{vx}{c^2}\right)$$

$$= 1.1547 \left(30 \times 10^{-6} \right.$$

$$\left. - \frac{1.5 \times 10^8 \times 10,000}{(3 \times 10^8)^2} \right)$$

41

$$t' = 1.1547(30 \times 10^{-6} - 16.667 \times 10^{-6})$$
$$= 15.396 \times 10^{-6}$$

The results from both transformations are listed in Table 3.3. It can be seen that for $v = 0.5c$ the two sets of result are significantly different.

Table 3.3 Space-time data from Galilean and Lorentz transformations

	S	S' Galilean	S' Lorentz
x	10,000 m	5,500 m	6,350.9 m
t	30 x 10^{-6} s	30 x 10^{-6} s	15.396 x 10^{-6} s

3.3 Time Dilation

An inertial system S' is moving at a speed v with respect to another inertial system S. Two events, which correspond to an observer reading the time from a stationary clock, occur at t'_1 and t'_2, respectively in S' as shown in Fig. 3.4. The time interval $(t'_2 - t'_1)$, denoted as T_0, is equal to the time duration read from the clock by the observer. T_0 is called the "proper time" as it corresponds to the time of a stationary clock.

In S, the two events occur at (x_1, t_1) and (x_2, t_2), respectively. Now the clock is moving at a speed v, and the space and time intervals of the two events are related by

$$x_2 - x_1 = v(t_2 - t_1) \qquad (3.3.1)$$

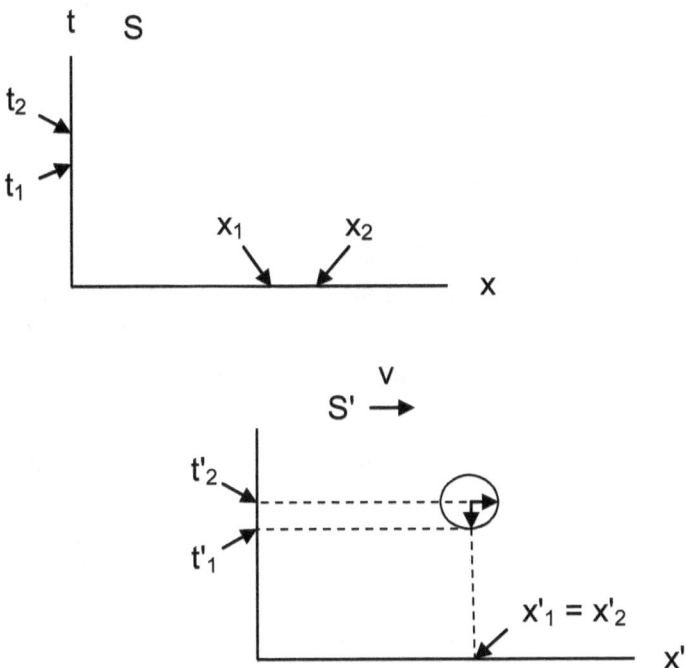

Fig. 3.4 Clock at rest in S'

According to the Lorentz transformations,

$$t'_2 - t'_1 = \gamma\left((t_2 - t_1) - \left(\frac{v}{c^2}\right)(x_2 - x_1)\right) \quad (3.3.2)$$

Combining Equations (3.3.1) and (3.3.2) we obtain

$$t'_2 - t'_1 = (t_2 - t_1)\sqrt{1 - \frac{v^2}{c^2}} \quad (3.3.3)$$

The quantity $(t_2 - t_1)$, denoted as T, is the time interval obtained by observers in S. Thus,

$$T = \frac{T_0}{\sqrt{1 - \frac{v^2}{c^2}}} \qquad (3.3.4)$$

$$T = \gamma T_0 \qquad (3.3.4)$$

T is always larger than T_0 and this phenomenon is called "time dilation". Since a stationary clock in S' is moving as viewed by observers in S, and smaller time interval corresponds to lower clock rate; time dilation is also referred to as "moving clocks run slow". For example, if the time interval read from a clock in an inertial system, in which the clock is at rest, is 1 second; then the same time interval measured in other inertial systems will be longer than 1 second.

An important point is that the time interval obtained in S' is the result of observation made by a single observer, whereas the time interval obtained in S is the result of observations made by two separate observers.

Since all inertial systems are equivalent, time dilation is reciprocal in that the time interval read from a stationary clock in S will be measured longer in S'. This raises a question: one of two identical clocks is stationary in S and another one is stationary in S', which one runs slower? We will get back to this subject in Chapter 6.

3.4 Space Contraction

A meter stick is placed along the x-axis and is at rest in S' which is moving at a speed v with respect to S. The two ends of the meter stick are at x'_1 and x'_2 as shown in Fig. 3.5. The space interval $(x'_2 - x'_1)$, denoted as L_0, is equal to the length of the meter measured in S'. L_0 is called the "proper length" as it corresponds to the length of a stationary object.

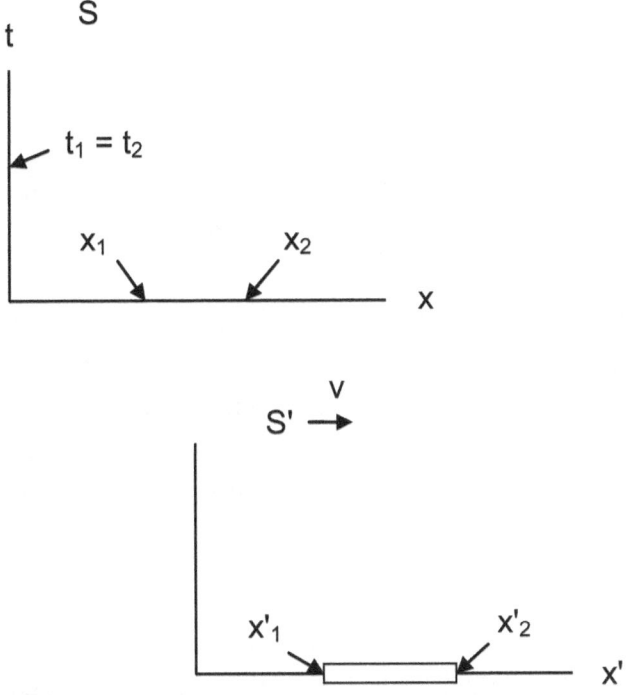

Fig. 3.5 Meter stick at rest in S'

Since the meter stick is moving in S; to measure its length, observers must mark the x-coordinates of the two ends, x_1 and x_2, simultaneously, i.e. at $t_1 = t_2$. The length L is then

$$L = x_2 - x_1$$

According to the Lorentz transformations,

$$x'_2 - x'_1 = \frac{x_2 - ct_2 - x_1 + ct_1}{\sqrt{1 - \frac{v^2}{c^2}}}$$

Since $t_1 = t_2$,

$$L_0 = \frac{L}{\sqrt{1 - \frac{v^2}{c^2}}} \qquad (3.4.1)$$

$$L = \frac{L_0}{\gamma} \qquad (3.4.2)$$

L is always smaller than L_0 and this phenomenon is called "space contraction". Since the meter stick is moving as viewed by observers in S, space contraction is also referred to as "moving rods contract". For example, a 1-meter long meter stick at rest in S' will be measured as shorter than 1 meter in S.

Like time dilation, space contraction is reciprocal in that the length of a meter stick at rest is S will be measured as shorter in S'. Now there is a question: one of two identical meters is stationary in S and another one is stationary in S', which one is actually shorter? We will get back to this subject in Chapter 6.

3.5 Simultaneity and Time Order of Events

A consequence of the Lorentz transformation is that two events that occur simultaneously in a system may not occur simultaneously in other systems. This can be deduced from the transformation equation for time interval, i.e.

$$\Delta t' = \gamma(\Delta t - \frac{v\Delta x}{c^2})$$

Since the time interval Δt in S is zero if the two events occur simultaneously; it can be seen from the above equation that the two events will also occur simultaneously in S', i.e. $\Delta t' = 0$, if they occur at the same space location in S, i.e. $\Delta x = 0$. Depending on the values of Δt and Δx, the time order of two events can be of one of the following cases:

Case 1: $\Delta t = \frac{v\Delta x}{c^2}$

Two events that are not simultaneous in S, i.e. $\Delta t \neq 0$, are simultaneous in S', i.e. $\Delta t' = 0$.

Case 2: $\Delta t > \frac{v\Delta x}{c^2}$

Time order of the two events is the same in S and S'.

Case 3: $\Delta t < \frac{v\Delta x}{c^2}$

Time order of the two events is not the same in S and S'.

The change of time order is due to the relative motion of two inertial systems. If two events are related causally, the time order will always be the same in all systems. We will get into more detail in the next section.

3.6 Causality and the Relativity

For two events that are related causally, there must be a signal or physical object traveling between two space locations. In the case that a person did something by the order of someone else, it is a signal that was traveling. In the case that a person was shot by a bullet from another person's gun, it is the bullet that was traveling. The speed v_s of a signal or physical object is related to the space and time intervals of the two events by

$$v_s = \frac{\Delta x}{\Delta t} \quad (3.6.1)$$

It has been shown that the condition for two events to change the time order is

$$\Delta t < \frac{v \Delta x}{c^2} \quad (3.6.2)$$

where v is the relative speed between two inertial systems. From Equations (3.6.1) and (3.6.2) we obtain

$$v v_s > c^2 \quad (3.6.3)$$

Equation (3.6.3) is the condition for two events, which are related causally, to change the time order. In physical world, nothing can travel faster than c. Therefore, this condition can never be satisfied, and the time order of two events related causally is the same in all inertial systems. Thus, the Relativity is in accord with the causality principle.

3.7 Space-time Interval

Space and time intervals between two events are not invariant under the Lorentz transformations, but there is an interval called the space-time interval that is invariant.

The space-time interval Δs in S is defined as

$$(\Delta s)^2 = (c\Delta t)^2 - ((\Delta x)^2 + (\Delta y)^2 + (\Delta z)^2) \quad (3.7.1)$$

The space-time interval $\Delta s'$ in S' is then

$$(\Delta s')^2 = (c\Delta t')^2 - ((\Delta x')^2 + (\Delta y')^2 + (\Delta z')^2) \quad (3.7.2)$$

The invariance of the space-time interval can be proved by first applying the Lorentz transformations on the right side of Equation (3.7.2) and obtain

$$c^2(\Delta t')^2 - ((\Delta x')^2 + (\Delta y')^2 + (\Delta z')^2)$$
$$= c^2\gamma^2 \left(\Delta t - \frac{v}{c^2}\Delta x\right)^2$$
$$- (\gamma^2(\Delta x - v\Delta t)^2 + (\Delta y)^2$$
$$+ (\Delta z)^2) \quad (3.7.3)$$

We next insert

$$\gamma = \frac{1}{\sqrt{1 - \frac{v^2}{c^2}}}$$

into Equation (3.7.3) and obtain

$$c^2(\Delta t')^2 - ((\Delta x')^2 + (\Delta y')^2 + (\Delta z')^2)$$
$$= (c\Delta t)^2 - ((\Delta x)^2 + (\Delta y)^2$$
$$+ (\Delta z)^2) \quad (3.7.4)$$

It can be seen from Equation (3.7.4) that

$$(\Delta s')^2 = (\Delta s)^2$$

The space-time interval is invariant in all inertial systems.

If the change of space coordinates is limited to the x-direction, then Equation (3.7.1) is reduced to

$$(\Delta s)^2 = (c\Delta t)^2 - (\Delta x)^2 \quad (3.7.5)$$

Under this circumstance, depending on the values of Δx and Δt, the relationship between two events can be divided into three cases.

Case 1: $c\Delta t = \Delta x$

The two events are called light-like. They are events describing light propagation between two space locations.

Case 2: $c\Delta t > \Delta x$

The two events are called time-like. They are events with short space interval or long time interval. For such events we can define the proper time $\Delta \tau$ as

$$\Delta \tau = \frac{\Delta s}{c} = \sqrt{(\Delta t)^2 - \left(\frac{\Delta x}{c}\right)^2} \quad (3.7.6)$$

Since Δs is an invariant in all inertial systems, $\Delta \tau$ is also an invariant. In a system S, if two events occur at the same space location, i.e. $\Delta x = 0$, Δt is equal to the time interval read from a clock at that location. It can be seen from Equation (3.7.6) that $\Delta \tau$ is equal to Δt. Thus, the proper time corresponds to the time read from a single clock at rest in S. In a different system S', the two events do not occur at the same location; however, Equation (3.7.6) allows us to find the proper time between the two events if the two events are time-like, and the value is equal to the time read from a single clock at rest in S. If two events are not time-like, Equation (3.7.6) will result in a value of zero or an imaginary number which is of course meaningless.

The two events in Table 3.3 are not time-like events since

$$c\Delta t = 3 \times 10^8 \times 30 \times 10^{-6} = 9,000$$

and
$$\Delta x = 10,000$$

Thus,
$$c\Delta t < \Delta x$$

If the space interval is changed from 10,000m to 6,000 m, then

$$c\Delta t > \Delta x$$

The two events now become time-like events and we can use Equation (3.7.6) to calculate the proper time of the time interval between the two events.

In system S,

$$\Delta\tau = \sqrt{(30 \times 10^{-6})^2 - \left(\frac{6000}{3 \times 10^8}\right)^2}$$

$$\Delta\tau = 22.36 \times 10^{-6} \ s$$

In a system S', which is moving in the x-direction at a speed of 0.5c with respect to S, the space and time intervals are

$$\Delta t' = 23.09 \times 10^{-6} \ s$$
$$\Delta x' = 1732.05 \ m$$

Thus,

$$\Delta\tau = \sqrt{(23.09 \times 10^{-6})^2 - \left(\frac{1732.05}{3 \times 10^8}\right)^2}$$
$$\Delta\tau = 22.36 \times 10^{-6} \ s$$

The proper time is indeed the same in both systems.

Let us use an example to elaborate. A passenger on a train read the time from his watch and the time is t_1, after a while the passenger read the time again and the time is t_2. The time duration T_0 read from the watch is then

$$T_0 = t_2 - t_1$$

In the train system S, the time interval T between the events of reading the time of the watch is equal to T_0, and the space interval is 0. In the ground system S', the time interval T' between the events is not equal to T_0, and the space interval is not 0. However, we can use the data in S' and Equation (3.7.6) to obtain the proper time between the events, and the proper time is equal to T_0.

Case 3: $c\Delta t < \Delta x$

The two events are called space-like. They are events with long space interval or short time interval. For such events we can define the proper distance (length) $\Delta\sigma$ as

$$\Delta\sigma = \sqrt{-(\Delta s)^2} = \sqrt{(\Delta x)^2 - (c\Delta t)^2} \qquad (3.7.7)$$

If two events are marking the space location of the two ends of an object simultaneously, i.e. $\Delta t = 0$, Δx is then the length of the object, and it can be seen from (3.7.7) that $\Delta\sigma$ is equal to Δx. Thus, $\Delta\sigma$ is the length of an object or the space separation (distance) between two space points. Equation (3.7.7) allows us to find the value of $\Delta\sigma$ using data obtained in any inertial system if the two events are space-like. If the two events are not space-like, Equation (3.7.7) will result in a value of zero or an imaginary number which is of course meaningless.

Let us use an example to elaborate. Passengers on a train measured the length of the train by marking the space locations of the two ends at the same time. The location of one end is at x_1 and another one at x_2. The length of the train L_0 is then

$$L_0 = x_2 - x_1$$

In the train system S, the space interval L between the events of marking the space locations of the two ends is equal to L_0, and the time interval is 0. In the ground system S', the space interval L' between the events is not equal to L_0, and the time interval is not 0. However, we can use the data in S' and Equation (3.7.7) to obtain the proper distance between the events, and the proper distance is equal to L₀.

3.8 Lorentz Transformations for Velocity

The Lorentz transformations for velocity can be derived from the Lorentz transformations for space and time coordinates. We first take the differential of Equations (3.1.16) - (3.1.19) and obtain

$$dx' = \frac{dx - vdt}{\sqrt{1 - \frac{v^2}{c^2}}} \quad (3.8.1)$$

$$dy' = dy \quad (3.8.2)$$

$$dz' = dz \quad (3.8.3)$$

$$dt' = \frac{dt - \left(\frac{v}{c^2}\right)dx}{\sqrt{1 - \frac{v^2}{c^2}}} \quad (3.8.4)$$

Dividing Equation (3.8.1) by Equation (3.8.4) we obtain

$$\dot{x}' = \frac{dx'}{dt'} = \frac{dx - vdt}{dt - \left(\frac{v}{c^2}\right)dx} \quad (3.8.5)$$

Dividing both the numerator and denominator of the right side of Equation (3.8.5) by dt we obtain

$$\dot{x}' = \frac{\dot{x} - v}{1 - \left(\frac{v}{c^2}\right)\dot{x}} \quad (3.8.6)$$

where

$$\dot{x} = \frac{dx}{dt}$$

Applying the above procedures to Equation (3.8.2) and Equation (3.8.3) respectively we obtain

$$\dot{y}' = \frac{\dot{y}\sqrt{1 - (\frac{v}{c})^2}}{1 - (\frac{v}{c^2})\dot{x}} \qquad (3.8.7)$$

$$\dot{z}' = \frac{\dot{z}\sqrt{1 - (\frac{v}{c})^2}}{1 - (\frac{v}{c^2})\dot{x}} \qquad (3.8.8)$$

Equations (3.8.6) - (3.8.8) are the Lorentz transformations for velocity.

3.9 Velocity Addition Law of the Relativity

From the Lorentz transformations for velocity we can obtain the velocity addition law of the Relativity.

Equation (3.8.6) is for transformation from S to S'. The transformation from S' to S is

$$\dot{x} = \frac{\dot{x}' + v}{1 + (\frac{v}{c^2})\dot{x}'} \qquad (3.9.1)$$

The equation can be written as

$$u = \frac{u' + v}{1 + \dfrac{u'v}{c^2}} \quad (3.9.2)$$

in which u and u' are the velocity of the object observed in S and S', respectively, and v is the velocity of S' relative to S. Equation (3.9.2) is thus the velocity addition law of the Relativity.

If a train moves at a speed v, and a light source on the train emits a light ray in the same direction of v; according to the velocity addition law of Newtonian physics, the speed of the light ray with respect to the platform is

$$u = c + v$$

The speed of the light ray with respect to the platform is greater than c which contradicts all experimental results. If the velocity addition law of the Relativity is used instead, then

$$u = \frac{c + v}{1 + \dfrac{cv}{c^2}} = c$$

The speed of the light ray with respect to the platform is still c. According to the velocity addition law of the Relativity, the speed of light is c in all inertial systems and the speed of any object can never exceed c.

3.10 Physical Speed and Relative Speed

The speed of an object we have discussed so far is the physical speed of the object, i.e. the speed of the object measured in a certain inertial system. All experimental results indicate that the physical speed of any object cannot exceed c.

Although the physical speed of an object cannot exceed c, the relative speed of two objects can.

In a lab, two particles are generated simultaneously. Particle A then moves at a speed of 0.7c in the x-direction and Particle B moves at a speed of 0.8c in the -x-direction, as shown in Fig. 3.6. The relative speed between the two particles is 1.5c which exceeds c.

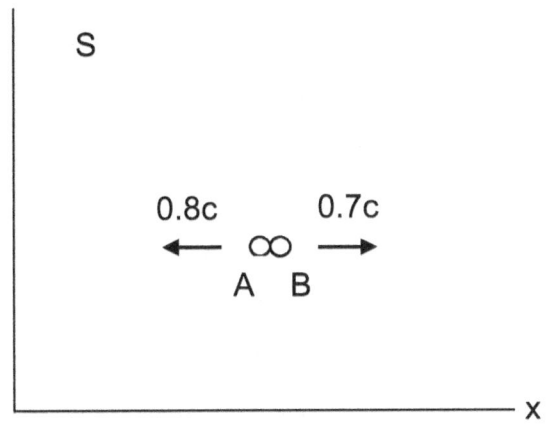

Fig. 3.6 Particle movement observed in the Lab system S

The above is an observation made in the lab system S. When observed in a system S', which is at rest with respect to Particle A, Particle B is then moving at a speed u' in the x-direction as shown in Fig. 3.7.

Using the Lorentz transformations we obtain

$$u' = 0.962c$$

Since Particle A is at rest, the relative speed of the two particles is now 0.962c. It is clear that the relative speed of two objects can be larger or smaller than c, whereas the physical speed of an object is always smaller than c.

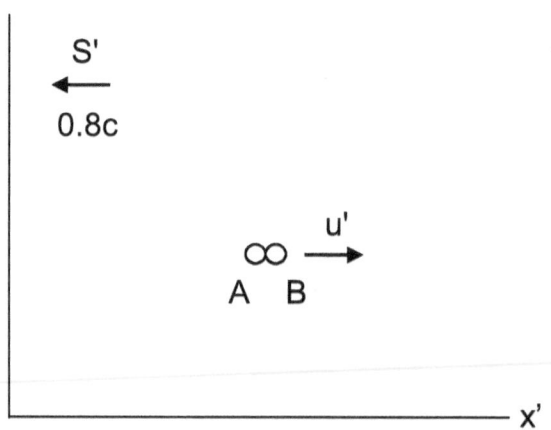

Fig. 3.7 Particle movement observed in S'

Chapter 4: Applications and Verifications of the Relativity

All new theories of science must be verified experimentally in order to be accepted and the Relativity is of no exception. In this chapter, we are going to give an account of several applications of the Relativity which also serve as experimental verifications of the theory.

4.1 Muon Decay

Many particles are unstable. They decay quickly into other particles after their production. The decay rate of an unstable particle is measured by its half-life T_h. If the original number of the particles is N, then after a time duration of T_h only approximately 1/2 N of them remain un-decayed. One of the unstable particles is the muon which has the same properties as the electron except that it is about 207 times heavier. The half-life of the muon measured in a system at rest with respect to the muon is 1.52×10^{-4} s.

In the universe there are high-energy particles called cosmic ray particles. When those particles enter the atmosphere of the earth and collide with nuclei in the atmosphere, muons are produced.

A group of muons flew at a speed of 0.98c from the sky toward the ground of the earth, the distance L they traveled in the time duration of a half-life is

$$L = 0.98 \times 3 \times 10^8 \times 1.52 \times 10^{-4} = 44{,}699 \text{ m}$$

If the number of muons measured in the sky 44,699m above the ground is N, the number of muons measured on the ground is supposed to be approximately 1/2 N; but the actual number is much larger, why?

In the analysis of a physical phenomenon in an inertial system, all quantities entering into the analysis must be obtained from that inertial system. In the above analysis, the half-life was obtained from a system at rest with respect to the muons, i.e, the muon system, whereas the distance the muons traveled was obtained from a system at rest with respect to the ground of the earth, i.e. the ground system.

To do the analysis of a physical phenomenon in an inertial system correctly, all space-time data obtained from other systems must be converted using the Lorentz transformations.

In the following sections, we are going to redo the analysis first in the ground system and then in the muon system.

4.1.1 Analysis in the Ground System

In the ground system, the muons were moving toward the ground as shown in Fig. 4.1. We must first convert the value of the half-life obtained in the muon system to that in the ground system. As the muon system moves at a speed of 0.98c with respect to the ground system,

$$\gamma = \frac{1}{\sqrt{1 - (0.98)^2}} = 5.025$$

Using the Lorentz transformations for time interval we obtain the half-life T_h in the ground system, and

$$T_h = \gamma T_{h0}$$

where T_{h0} is the half-life measured in the muon system. The value of T_h is

$$T_h = 7.638 \times 10^{-4} \ s$$

It takes the muons 1.52×10^{-4} s, only about 1/5 of the half-life, to reach the ground; thus, the number of un-decayed muons is thus much larger than half of the original number.

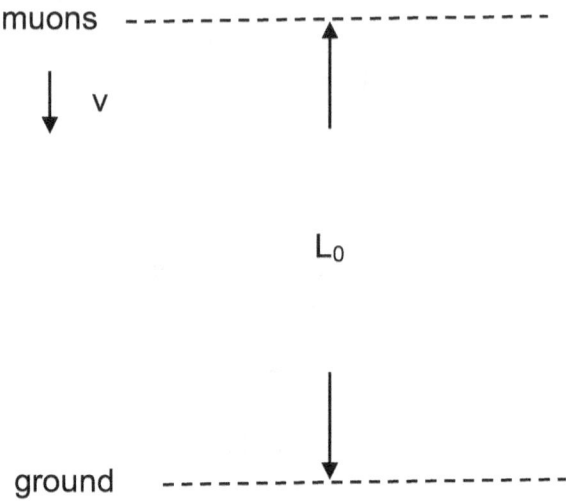

Fig. 4.1 Analysis in the ground system

4.1.2 Analysis in the Muon System

In the muon system, the ground was moving at a speed of 0.98c toward the muons as shown in Fig. 4.2. This time we must first convert the value of the distance obtained in the ground system to that in the muon system.

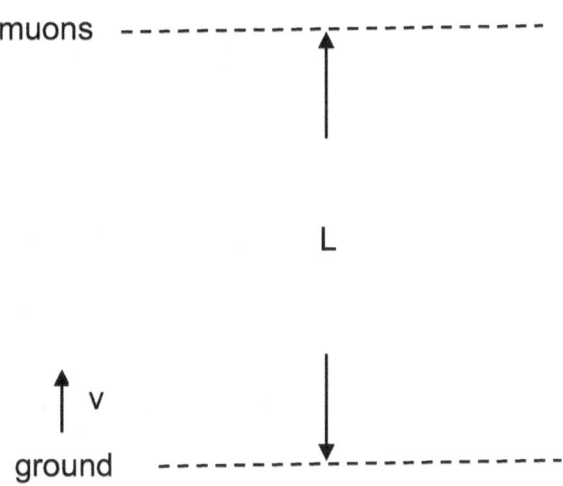

Fig. 4.2 Analysis in the muon system

Using the Lorentz transformations for space interval we obtain the distance L in the muon system, and

$$L = \frac{L_0}{\gamma}$$

$$L = \frac{44,699}{5.025} = 8,895 \; m$$

The time duration required for the ground to travel a distance of 8,895 m is then

$$T = \frac{8,895}{0.98 \times 3 \times 10^8} = 0.3026 \times 10^{-4} s = 0.198 T_{h0}$$

The time duration is only about 1/5 of the half-life of the muon; the number of un-decayed muons is thus much larger than half of the original number.

4.2 Stellar Aberration

If a person wants to observe a star directly overhead using a telescope as shown in the left side of Fig. 4.3, because of the motion of the earth, he or she has to tilt the telescope by a small angle as shown in the right side of Fig. 4.3; otherwise the light from the star would hit the wall of the telescope and cannot reach the eye. Although the angle is extremely small, it has been observed by astronomers.

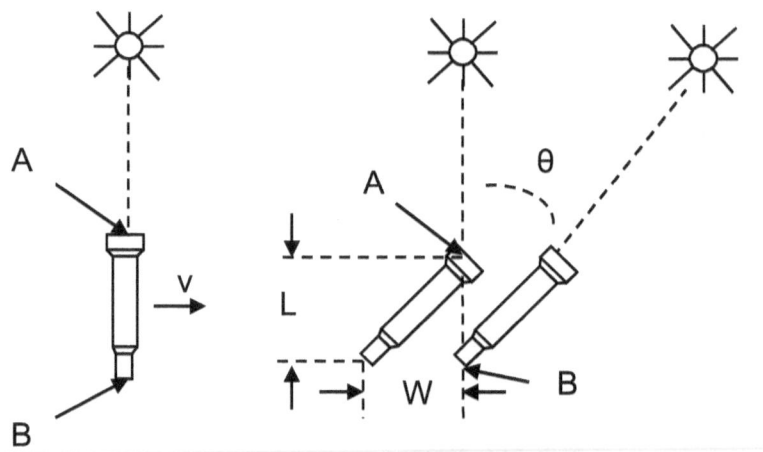

Fig. 4.3 Stellar aberration

Since the telescope is tilted and is moving to the right, the starlight does not hit the wall of the telescope and continues travelling down. When the starlight reaches the point B, the bottom end of the telescope also reaches that point. The time duration for the starlight to travel from the point A to the point B is T, and the distance that the starlight and telescope travelled in this time duration is L and W, respectively. In conventional physics, the angle is calculated as the following:

$$\tan\theta = \frac{W}{L} = \frac{vT}{cT} = \frac{v}{c} \qquad (4.2.1)$$

where v is the orbital speed of the earth around the sun. Inserting the values of v and c into Equation (4.2.1) we obtain

$$\tan\theta = \frac{3 \times 10^4}{3 \times 10^8}$$

$$\theta = 5.73 \times 10^{-3} \ radian$$

$$\theta = 20.5''$$

The value of θ is 20.5" (1 radian = 3600") and is in agreement with that observed by astronomers.

The stellar aberration can be regarded as a result of observations made in different inertial systems. Let S be the system at rest with respect to the earth and S' be the system at rest with respect to the sun. Recall that the starlight is actually the sunlight reflected by the star. Because of the relative motion between the sun and the earth, S' is moving at a speed v in the x-direction with respect to S. When observed in S' the starlight travels along a vertical line as shown in Fig. 4.4. When observed in S the starlight travels along a line that makes an angle θ with the horizontal as shown in Fig. 4.5. We are going to find out the value of θ using the Lorentz transformations.

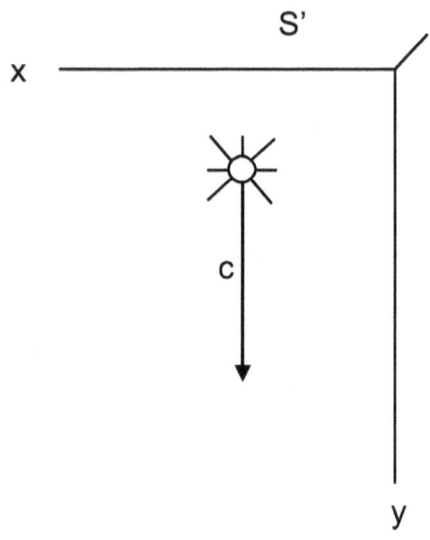

Fig. 4.4 Starlight observed in the sun system

In S', the speed of the starlight has y-component only, and

$$u'_y = c$$

$$u'_x = 0$$

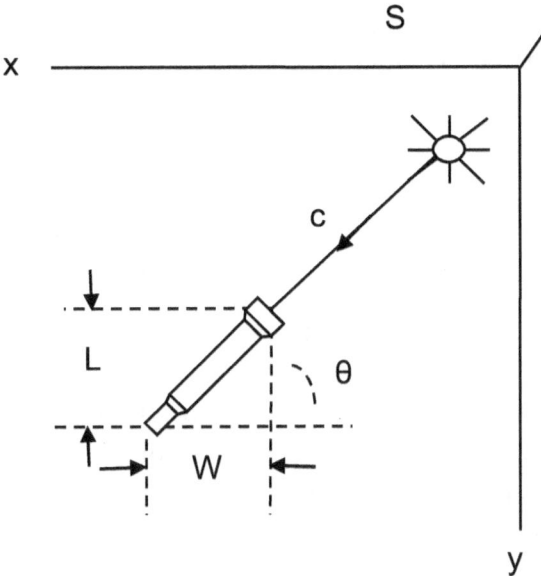

Fig. 4.5 Starlight observed in the earth system

In S, according to the Lorentz transformations, the velocity components are

$$u_x = v \quad (4.2.2)$$

$$u_y = c \sqrt{1 - \frac{v^2}{c^2}} \quad (4.2.3)$$

where v is the relative speed between S and S' and is the orbital speed of the earth around the sun.

67

It can be seen From Fig. 4.5 that

$$\frac{W}{L} = \frac{u_x}{u_y} = \frac{v}{c\sqrt{1 - \frac{v^2}{c^2}}} \qquad (4.2.4)$$

As v is extremely small as compared to c,

$$\frac{v^2}{c^2} \approx 0$$

Thus,

$$\tan \theta = \frac{W}{L} = \frac{v}{c} \qquad (4.2.5)$$

Equation (4.2.5) is the same as Equation (4.2.1). Thus, the expression in conventional physics is only an approximation to the accurate one, i.e. the expression in the Relativity.

4.3 Doppler Effect

When there is a relative motion between a sound source and an observer, the observed sound frequency depends on the speed of the relative motion. This phenomenon is called the Doppler Effect. The Doppler Effect also occurs on light signals.

Sounds result from variation of the pressure of the air, and the frequency represents the number of variation per unit time. If a source produces a sound with a frequency of 5,000 per second as shown in Fig. 4.6, an observer at rest with respect to the source will sense 5.000 variations per second. In other words, the sound frequency observed by the observer is 5,000/s. However, if the observer is moving toward the sound source, he or she will sense 5.000 variations in less than one second. Thus, the frequency observed by the observer is higher than 5,000/s. If the observer is moving away from the sound source, he or she will sense less than 5.000 variations in one second. Thus, the frequency observed by the observer is lower than 5,000/s.

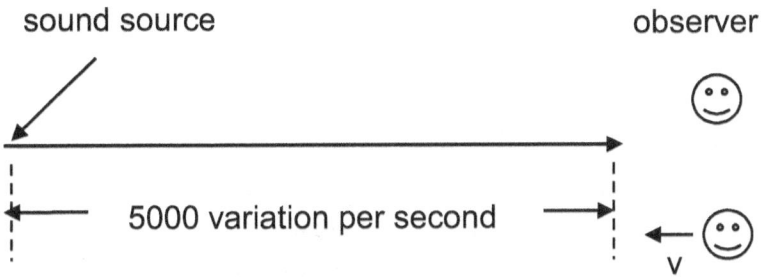

Fig. 4.6 Doppler Effect of sound wave

In conventional physics, the expressions for Doppler Effect can be derived as the following. A light source is emitting a train of light pulses of period t_s as shown in Fig. 4.7. The frequency of the light pulses is f_s which by definition is equal to $1/t_s$.

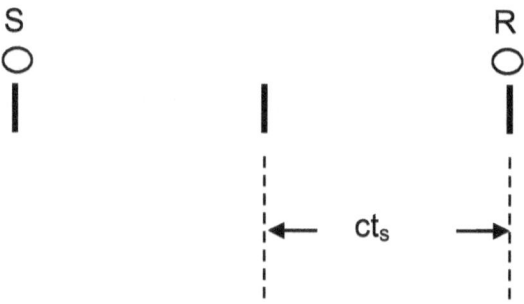

Fig. 4.7 Observer at rest

If the observer is at rest with respect to the source; when he or she receives a pulse, the next pulse will reach him or her after a time interval of t_s. However, if the observer is moving away from the source at a speed v, when he or she receives a pulse, the next pulse has to travel an extra distance and to take an extra time to reach him or her. As a result, the actual period and frequency of the signal sensed by the observer are t_{rs} and f_{rs}, respectively. The situation is shown in Fig. 4.8.

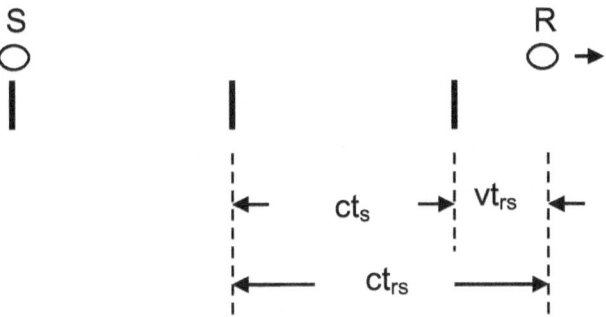

Fig. 4.8 Observer moving away from source

It can be seen from Fig. 4.8 that

$$ct_{rs} = ct_s + vt_{rs}$$

$$\frac{c}{f_{rs}} = \frac{c}{f_s} + \frac{v}{f_{rs}} \quad (4.3.1)$$

where c is the speed of light. Equation (4.3.1) can be written as

$$f_{rs} = (1 - \beta)f_s \quad (4.3.2)$$

where

$$\beta = \frac{v}{c}$$

If the observer is moving toward the source, the expression can be obtained in a similar way and the result is

$$f_{rs} = (1 + \beta)f_s \quad (4.3.3)$$

The Doppler Effect can be regarded as a result of observations made in different inertial systems. Thus, it can be derived using the Lorentz transformations.

Since t_s is the period measured in a system at rest with respect to the light source and t_{rs} is the period measured in a system moving with a speed v with respect to the light source,

$$t_{rs} = Y t_s$$

$$Y = \frac{1}{\sqrt{1 - \beta^2}}$$

Using the same approach as that in deriving Equation (4.3.1) we obtain

$$f_{rs} = f_s \frac{\sqrt{1 - \beta^2}}{1 + \beta} = f_s \sqrt{\frac{1 - \beta}{1 + \beta}} \qquad (4.3.4)$$

If the observer is moving toward the source, the equation can be obtained in a similar way and the result is

$$f_{rs} = f_s \sqrt{\frac{1 + \beta}{1 - \beta}} \qquad (4.3.5)$$

Equation (4.3.4) can be written as

$$f_{rs} = f_s (1 - \beta)^{\frac{1}{2}} (1 + \beta)^{-\frac{1}{2}} \qquad (4.3.6)$$

Applying the binominal theorem to $(1 - \beta)^{\frac{1}{2}}$ and $(1 + \beta)^{-\frac{1}{2}}$, Equation(4.3.6) becomes

$$f_{rs} = f_s \left(1 - \beta + \frac{1}{2}\beta^2 + \cdots\cdots\right) \qquad (4.3.7)$$

Similarly, Equation (4.3.5) becomes

$$f_{rs} = f_s \left(1 + \beta + \frac{1}{2}\beta^2 + \cdots\cdots\right) \qquad (4.3.8)$$

When the relative motion between the source and observer is slow, the value of β^2 is extremely small; terms other than the first two in the bracket can be ignored and the equations become

$$f_{rs} = f_s(1 - \beta) \qquad (4.3.9)$$

$$f_{rs} = f_s(1 + \beta) \qquad (4.3.10)$$

Equations (4.3.9) and (4.3.10) are the same as those in conventional physics. Thus, the expressions in conventional physics are only approximations to the accurate ones, i.e. the expressions in the Relativity.

Chapter 5: Dynamics of the Relativity

The kinematics of the Relativity was discussed in Chapter 3. In this chapter, we are going to discuss the dynamics of the Relativity. Subjects to be covered include momentum, energy, conservation of momentum and energy, equivalence of mass and energy, etc.

5.1 Momentum

In Newtonian physics, the momentum of an object is defined as

$$p = mu \quad (5.1.1)$$

where p is momentum, m is mass, and u is speed. The mass of an object is an invariant.

Is the above definition of momentum valid also in the Relativity? We are going to examine using a simple case in which two identical objects of mass m are moving toward each other with the same speed u as shown in Fig. 5.1a. After the collision, the two objects form a composite at rest. The rest mass of the composite is M.

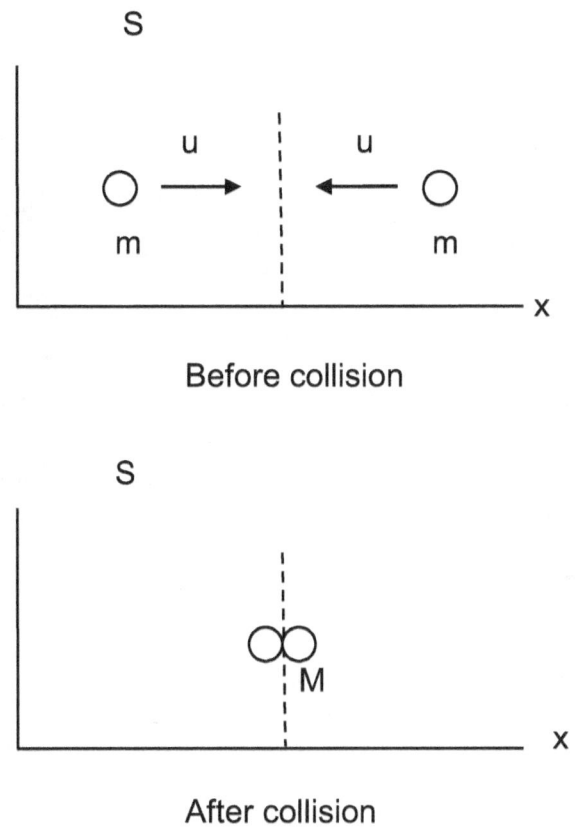

Fig. 5.1a Collision observed in the lab system

We first observe the collision in the lab system S. In Newtonian physics, the momentum of the system before collision is

$$p_{before} = mu - mu = 0 \quad (5.1.2)$$

After the collision, because of mass conservation

$$M = m + m = 2m \quad (5.1.3)$$

The momentum of the system is then

$$p_{after} = (2m)0 = 0 \quad (5.1.4)$$

The momentum is indeed conserved. The analysis in the Relativity is the same as that of Newtonian physics, and the momentum is also conserved.

We next observe in a system S' which is moving at a speed of $v = u$ in the -x-direction with respect to the lab system S. The situation in Newtonian physics is shown in Fig. 5.1b. According to the Galilean transformations for velocity, the object on the right side is at rest, the object on the left side is moving to the right with a speed 2u, and the composite is moving to the right with a speed u.

The momentum of the system before collision is

$$p_{before} = m\,(2u) = 2mu$$

The momentum of the system after collision is

$$p_{after} = (2m)u = 2mu$$

The momentum is still conserved.

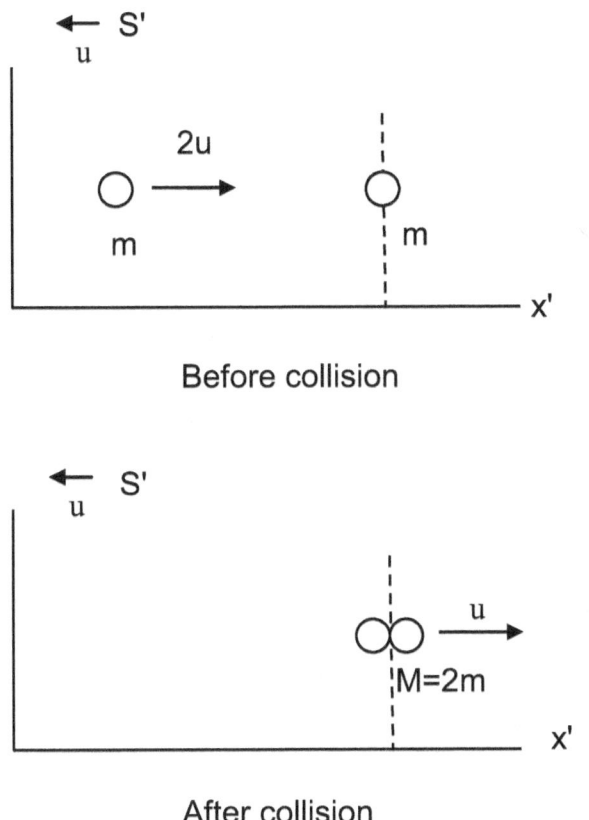

Fig. 5.1b Collision observed in S' in Newtonian physics

The situation in the Relativity is different from that in Newtonian physics. Now the object on the right side is at rest, the object on the left side is moving to the right with a speed u'_1, and the composite is moving to the right with a speed u'_2. The situation is shown in Fig. 5.2.

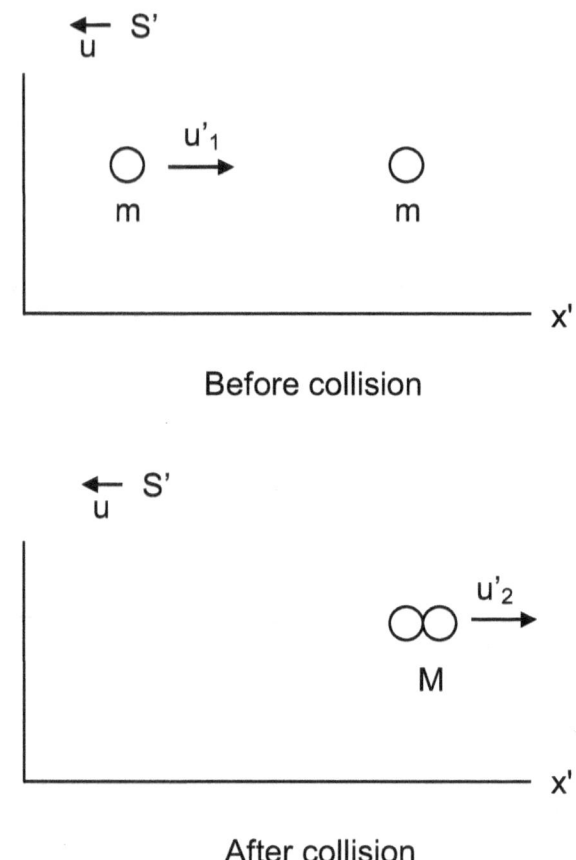

Before collision

After collision

Fig. 5.2 Collision observed in S' in the Relativity

According to the Lorentz transformations for velocity,

$$u'_1 = \frac{u - v}{1 - \frac{uv}{c^2}}$$

$$u'_2 = \frac{0 - v}{1 - \frac{0v}{c^2}}$$

Since v = -u,

$$u'_1 = \frac{u + u}{1 + \frac{uu}{c^2}} \qquad (5.1.5a)$$

$$u'_2 = u \qquad (5.1.5b)$$

The momentum before collision is

$$P_{before} = mu'_1 = \frac{2mu}{1 + \frac{u^2}{c^2}} \qquad (5.1.6)$$

The momentum after collision is

$$P_{after} = Mu'_2 = 2mu \qquad (5.1.7)$$

It is obvious that

$$P_{before} \neq P_{after}$$

If we stick to the definition of momentum of Newtonian physics, momentum conservation is violated.

It was found that momentum conservation is not violated if the momentum is defined as

$$p = m(u)u \qquad (5.1.8)$$

The mass of an object is no longer a constant, and is instead a function of the speed of the object. The mass of an object at rest is called the ''rest mass'' and is designated as m_0.

The mass of a moving object is called the "relativistic mass" or simply the mass and is designated as m(u) or m.

We need to find out how the mass of an object varies with the speed of the object, or equivalently the relationship between m and m_0. The derivation is lengthy and somewhat complicated. Those who are not interested in the mathematical detail can either skip or glance through and just pay attention to the result.

We consider a case in which an object is moving to the right and collides with another object that is of the same rest mass and is originally at rest. After the collision, the two objects form a composite moving to the right. The situation observed in the lab system S is shown in Fig. 5.3.

When the collision is observed in the lab system S; before the collision, one of the objects is moving and one is at rest, and

$$m_1 = m(u_1)$$

$$m_2 = m(0) = m_0$$

When the collision is observed in a system S' moving at a speed v, which is deliberately chosen to be equal to u_1, in the x-direction with respect to the lab system S, the situation becomes that shown in Fig. 5.4.

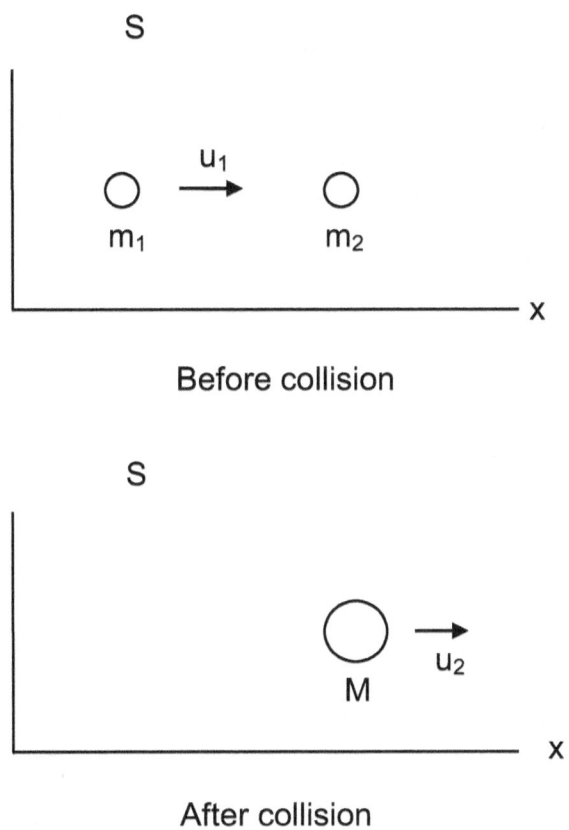

Fig. 5.3 Collision observed in S

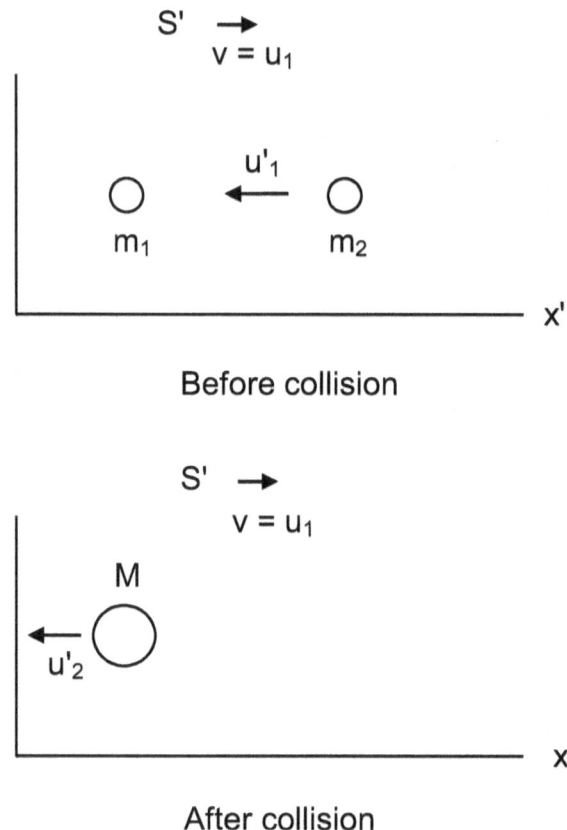

Before collision

After collision

Fig. 5.4 Collision observed in S'

In S, conservation of mass requires that

$$M(u_2) = m(u_1) + m(0) \quad (5.1.9)$$

In S', conservation of momentum requires that

$$M(u'_2)u'_2 = m(u'_1)u'_1 \quad (5.1.10)$$

Since $v = u_1$ and $m_1(0) = m_2(0)$; collisions in S and S' are symmetrical, and we can conclude that

$$u'_1 = -u_1$$

$$u'_2 = -u_2$$

Thus, Equation (5.1.10) can be written as

$$M(u_2)u'_2 = m(u_1)u_1 \quad (5.1.11)$$

Inserting Equation (5.1.9) into Equation (5.1.11) we obtain

$$\big(m(u_1) + m(0)\big)u'_2 = m(u_1)u_1 \quad (5.1.12)$$

$$m(u_1) = m(0)\frac{u'_2}{u_1 - u'_2} \quad (5.1.13)$$

According to the law of velocity addition,

$$u' = \frac{u + v}{1 + \dfrac{uv}{c^2}}$$

In our case, we have

$$u' = u'_2$$

$$u = u_2 = -u'_2$$

$$v = u_1$$

Thus,

$$u'_2 = \frac{-u'_2 + u_1}{1 - \frac{u'_2 u_1}{c^2}} \qquad (5.1.14)$$

Solving (5.1.14) for u_1 we obtain

$$u_1 = \frac{2u'_2}{1 + \frac{(u'_2)^2}{c^2}} \qquad (5.1.15)$$

From Equation (5.1.15) we obtain

$$\frac{u'_2}{u_1 - u'_2} = \frac{1 + \frac{(u'_2)^2}{c^2}}{1 - \frac{(u'_2)^2}{c^2}} \qquad (5.1.16)$$

Since

$$(1 - \frac{(u'_2)^2}{c^2})^2 = (1 + \frac{(u'_2)^2}{c^2})^2 - 4\frac{(u'_2)^2}{c^2}$$

thus,

$$\left(\frac{1 - \frac{(u'_2)^2}{c^2}}{1 + \frac{(u'_2)^2}{c^2}}\right)^2 = 1 - \frac{4\frac{(u'_2)^2}{c^2}}{\left(1 + \frac{(u'_2)^2}{c^2}\right)^2} \qquad (5.1.17)$$

Inserting Equation (5.1.15) into Equation (5.1.17) we obtain

$$\left(\frac{1 - \frac{(u'_2)^2}{c^2}}{1 + \frac{(u'_2)^2}{c^2}}\right)^2 = 1 - \frac{(u_1)^2}{c^2} \qquad (5.1.18)$$

Comparing Equation (5.1.18) with Equation (5.1.16) we obtain

$$\frac{u'_2}{u_1 - u'_2} = \frac{1}{\sqrt{1 - \frac{(u_1)^2}{c^2}}} \qquad (5.1.19)$$

Equation (5.1.13) can now be written as

$$m(u_1) = m(0)\frac{1}{\sqrt{1 - \frac{(u_1)^2}{c^2}}} \qquad (5.1.20)$$

Thus, the relationship between m(u) and m(0) is

$$m(u) = m(0)\frac{1}{\sqrt{1 - \frac{(u)^2}{c^2}}} \qquad (5.1.21)$$

where u is the speed of the object.

Equation (5.1.21) can be written as

$$m = \gamma m_0$$

where

$$\gamma = \frac{1}{\sqrt{1 - \frac{u^2}{c^2}}} \qquad (5.1.22)$$

At first glance, it looks that the ''γ'' in Equation (5.1.22) is the same as that of the Lorentz transformations. However, they are very different. The speed in Equation (5.1.22) is the speed of an object observed in an inertial system, whereas the speed in the Lorentz transformations is the relative speed of two inertial systems.

5.2 Kinetic Energy

In Newtonian physics, the kinetic energy K of an object moving at a speed u is defined as the work done by a force to raise the speed of the object from zero to u. If the force and the motion of the object are both limited to the x-direction, then

$$K = \int_{u=0}^{u=u} F dx \quad (5.2.1)$$

Since

$$F = m_0 a \quad (5.2.2)$$

thus,

$$K = \int m_0 a dx$$

As acceleration is the time derivative of velocity, Equation (5.2.1) can be written as

$$K = \int m_0 \frac{du}{dt} dx$$

$$K = \int m_0 \frac{dx}{dt} du$$

The kinetic energy of an object at speed u is then

$$K = m_0 \int_0^u u du$$

86

$$K = \frac{1}{2} m_0 u^2 \qquad (5.2.3)$$

Equation (5.2.3) is the expression for kinetic energy in Newtonian physics.

In the Relativity, since mass is not a constant, Equation (5.2.2) is not a valid definition of force and must be replaced with the general definition

$$F = \frac{dp}{dt} = \frac{d}{dt}(mu) \qquad (5.2.4)$$

If the directions of force and motion are both limited to the x-direction, then

$$K = \int F dx \qquad (5.2.5)$$

Inserting Equation (5.2.4) into Equation (5.2.5) we obtain

$$K = \int \frac{d}{dt}(mu) dx$$

$$K = \int d(mu) \frac{dx}{dt}$$

$$K = \int d(mu) u$$

$$K = \int (mdu + udm) u$$

$$K = \int (mudu + u^2 dm) \qquad (5.2.6)$$

We can obtain the following relation from Equation (5.1.21)

$$m^2c^2 - m^2u^2 = m_0{}^2c^2$$

Since the right side of the above equation is a constant, the following can be obtained by taking the differential of the equation

$$2mc^2dm - m^2 2udu - u^2 2mdm = 0$$

$$mudu + u^2dm = c^2dm \quad (5.2.7)$$

Inserting Equation (5.2.7) into Equation (5.2.6) and taking into account that m_0 corresponds to $u = 0$ and m corresponds to $u = u$ we obtain

$$K = \int_{m_0}^{m} c^2 dm$$

$$K = mc^2 - m_0c^2$$

$$K = m_0c^2(\gamma - 1) \quad (5.2.8)$$

Equation (5.2.8) is the expression for kinetic energy in the Relativity.

Equation (5.2.8) can be written as

$$K = m_0c^2\left((1 - \beta^2)^{-\frac{1}{2}} - 1\right) \quad (5.2.9)$$

where

$$\beta = \frac{u}{c} \quad (5.2.10)$$

Using the binominal theorem we can obtain

$$K = m_0c^2 \left(1 + \frac{1}{2}\beta^2 + \frac{3}{8}\beta^4 + \cdots - 1\right)$$

$$K = m_0c^2 \cdot \frac{1}{2}\beta^2 \left(1 + \frac{\frac{3}{8}\beta^4}{\frac{1}{2}\beta^2} + \cdots\right)$$

$$K = \frac{1}{2}m_0u^2 \left(1 + \frac{3}{4}\beta^2 + \cdots\right) \quad (5.2.11)$$

When the speed of the object is very low, all terms except the first one in the bracket are negligible, and Equation (5.2.11) becomes

$$K \cong \frac{1}{2}m_0u^2$$

This indicates that the expression for kinetic energy in Newtonian physics is an approximation at low speeds. When the speed approaches the speed of light, Equation (5.2.8) must be used.

We can regard ''m_0c^2'' as the rest energy of an object and define the total energy E of an object as

$$E = mc^2 \quad (5.2.12)$$

Equation (5.2.8) can be written as

$$E = m_0c^2 + K \quad (5.2.13)$$

Equation (5.2.13) states that the total energy of an object is the sum of the rest energy and kinetic energy.

5.3 Equivalence of Mass and Energy

The mass of an object increases with the speed of the object. The question is: where does the additional amount of mass come from? To answer this question, we need to reexamine the principle of mass conservation and the principle of energy conservation.

5.3.1 Conservation of Mass and Conservation of Energy

Two identical objects are moving toward each other with the same speed u as shown in Fig. 5.5. The mass of each object is m. After the collision, the objects form a composite at rest. The mass of the composite is M_0.

We already know that

$$E = mc^2 \qquad (5.3.1)$$

or equivalently

$$E = m_0 c^2 + K \qquad (5.3.2)$$

We also know that

$$m = \gamma m_0 \qquad (5.3.3)$$

and

$$\gamma = \frac{1}{\sqrt{1 - \dfrac{u^2}{c^2}}} \qquad (5.3.4)$$

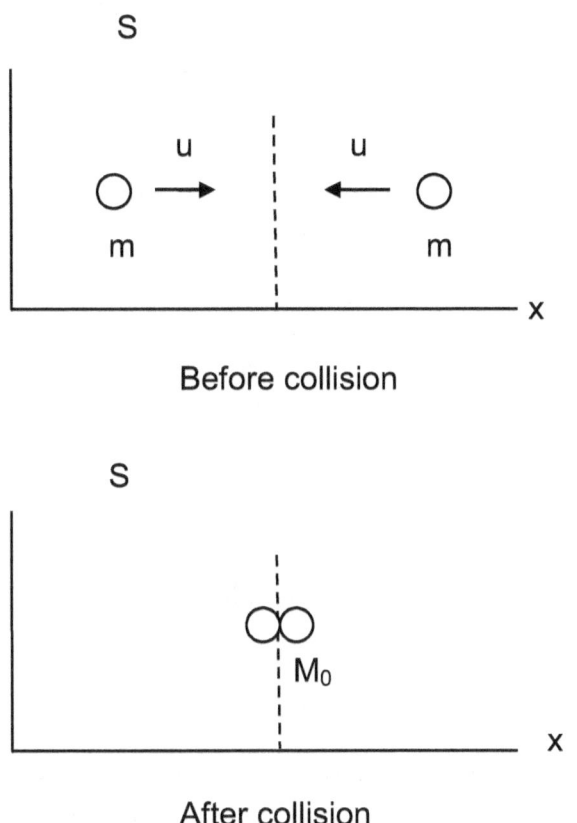

Fig. 5.5 Inelastic collision of two objects

Before the collision, the speed of each object is u, and the mass of the system is 2m. After the collision, the mass of the system is M_0. Since mass must be conserved in the collision,

$$M_0 = 2m = 2\gamma m_0 \qquad (5.3.5)$$

Before the collision, the rest mass of the system is $2m_0$, whereas after the collision, the rest mass of the system is $2\gamma m_0$. The difference is

$$\Delta M_0 = 2\gamma m_0 - 2m_0 = 2m_0(\gamma - 1) \qquad (5.3.6)$$

According to Equation (5.2.8), the kinetic energy of each object before the collision is

$$K = m_0(\gamma - 1)c^2 \qquad (5.3.7)$$

The kinetic energy of the system is $2m_0(\gamma - 1)c^2$ before the collision, and is zero after the collision. The difference is

$$\Delta K = 2m_0(\gamma - 1)c^2 \qquad (5.3.8)$$

Comparing Equation (5.3.6) with Equation (5.3.8) we obtain

$$\Delta K = \Delta M_0 c^2 \qquad (5.3.9)$$

This is an indication that some kinetic energy is converted into rest mass, and is also an indication that energy and mass are equivalent and exchangeable.

The equivalence of mass and energy was first proposed by Einstein with his famous equation

$$E = mc^2 \qquad (5.3.10)$$

As mass is equivalent to energy, it can be measured with the unit of energy. In modern physics, ``eV/c^2'' is often used as the unit of mass. The "eV" is an abbreviation for electron-Volt, a unit of energy.

The relation between "eV/c²" and Kg is

$$1\frac{eV}{c^2} = 1.78 \times 10^{-36} \text{ Kg} \quad (5.3.11)$$

The mass of an electron is 9.11×10^{-31} Kg. It can be expressed as 0.511 MeV/c² ($1\ M = 1 \times 10^6$), or simply 0.511 MeV with the "/c²" dropped for succinctness.

Since the value of c is extremely large, Equation (5.3.10) indicates that a small amount of mass can be converted into a huge amount of energy. This is how nuclear power is generated.

5.3.2 Conversion between Mass and Energy

The rest energy of a system consists of the masses of the constituent particles and the energy that binds the constituent particles together.

There are two types of systems formed by a group of particles: unbound and bound. In an unbound system, particles are not bound and can leave the system anytime. An example of unbound system is a group of photons in a light ray. In a bound system, particles are bound together by a binding force. Particles in a bound system cannot break away from the system unless an external energy, which is strong enough to overcome the binding energy, is applied. An example of bound system is the hydrogen atom. A hydrogen atom is formed by a nucleus and an electron as shown in Fig. 5.6. A binding energy of 13.6 eV exists between the nucleus and the electron.

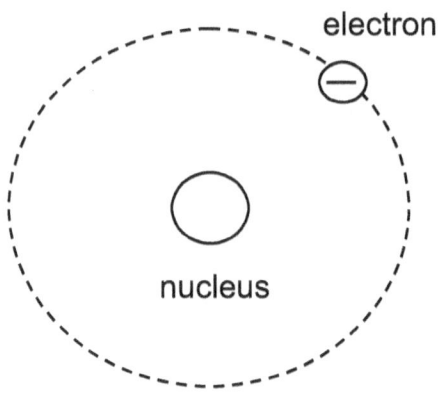

Fig. 5.6 Hydrogen atom

Most nuclei contain several protons and neutrons, but the nucleus of a hydrogen atom contains only a proton. Thus, the constituents of the hydrogen atom are a proton and an electron. The rest energy of a hydrogen atom is the sum of the mass of a proton, the mass of an electron, and the binding energy between the nucleus and the electron. If an external energy, which is equals to or larger than the binding energy, is applied to a hydrogen atom, the atom is broken up into a hydrogen nucleus and an electron.

When an electron and a proton combine to form a hydrogen atom; some energy, which is equal to the binding energy of the hydrogen atom, is released in the form of electromagnetic radiation. The binding energy is extremely small as compared to the total energy of a hydrogen atom, and is normally ignored in chemical reactions. However, the binding energy between nucleons, i.e. protons and neutrons, is much larger.

The nucleus of a deuterium is called the deuteron. The deuteron consists of a proton and a neutron as shown in Fig. 5.7. The masses of the proton and neutron are

$$m_p = 939.371 \text{ MeV}$$

$$m_n = 940.669 \; MeV$$

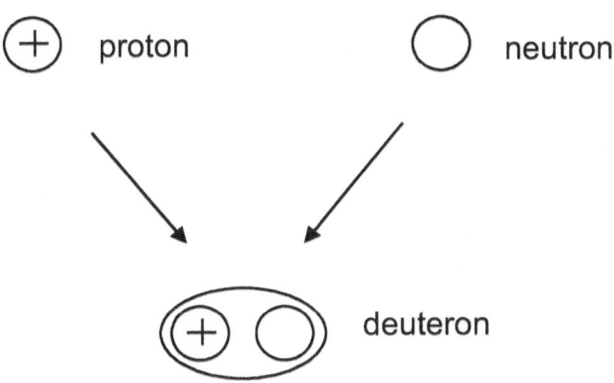

Fig. 5.7 Proton and neutron combine to form a deuteron

The sum is 1,880.039 MeV. The mass of a deuteron is 1,877,805 MeV. In the process of forming a deuteron, an amount of 2.23 MeV is released in the form of electromagnetic radiation. Although this is still a small amount of energy; when a large number of deuterons are formed at the same time, a huge amount of energy can be released.

The above is about conversion of a portion of the mass of an object into energy. The mass of an object can be completely converted into energy. Conversely, energy can be converted into mass and new particles are thus generated. Two well known examples of conversion between mass and energy are the positron-electron pair creation and positron-electron pair annihilation.

The positron and electron are antiparticle to each other. They have the same properties except that the positron carries a unit of positive charge whereas the electron carries a unit of negative charge. When a positron and electron meet, they annihilate each other and generate two or more photons as shown in Fig. 5.8. Since the photon has no mass, the process is a conversion of mass to energy. In the process, the mass of the positron and electron is converted into energy completely. A reverse process is the positron-electron pair creation in which a photon relinquishes its energy and creates a positron and an electron, as shown in Fig. 5.9.

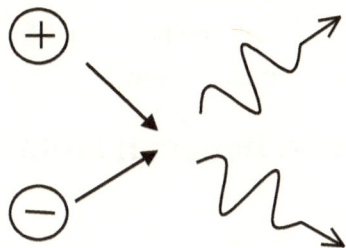

Fig. 5.8 Positron-electron pair annihilation

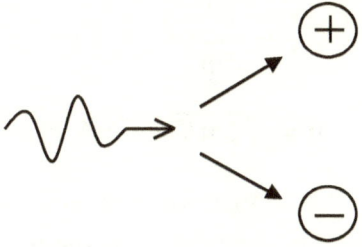

Fig. 5.9 Positron-electron pair creation

5.3.3 New Law of Energy and Mass Conservation

Since energy can be converted into mass and vice versa, energy and mass are not separately conserved. Instead, total energy is conserved, and the new law of conservation is called mass-energy conservation.

5.4 Connection between Momentum and Energy

In Newtonian physics, the connection between momentum p and kinetic energy K can be derived as the following.

$$K = \frac{1}{2}mu^2$$

$$K = \frac{1}{2m}(mu)^2$$

$$K = \frac{p^2}{2m} \quad (5.4.1)$$

$$p = \sqrt{2mK} \quad (5.4.2)$$

In the Relativity, the connection between momentum and kinetic energy must be derived from the new expression for momentum.

$$p = \gamma m_0 u$$

$$p^2c^2 = \gamma^2 m_0{}^2 u^2 c^2$$

$$p^2c^2 = \gamma^2 (m_0 c^2)^2 \left(\frac{u}{c}\right)^2 \quad (5.4.3)$$

We know that

$$\gamma = \frac{1}{\sqrt{1 - \left(\frac{u}{c}\right)^2}}$$

Thus,

$$\frac{1}{\gamma} = \sqrt{1 - \left(\frac{u}{c}\right)^2}$$

and

$$\left(\frac{u}{c}\right)^2 = 1 - \frac{1}{\gamma^2} \quad (5.4.4)$$

Inserting Equation (5.4.4) into Equation (5.4.3) and taking into consideration that "m_0c^2" is the rest energy E_0 we obtain

$$p^2c^2 = \gamma^2 E_0{}^2 \left(1 - \frac{1}{\gamma^2}\right) \quad (5.4.5)$$

We can reorganize Equation (5.4. 5) and obtain

$$p^2c^2 = \gamma^2 E_0{}^2 - E_0{}^2$$

$$p^2c^2 = E^2 - E_0{}^2$$

$$E^2 = E_0{}^2 + (pc)^2 \quad (5.4.6)$$

Equation (5.4.6) is the connection between total energy, rest energy, and momentum. It plays an important role in modern physics. As an example we are going to use it to specify the momentum, kinetic energy, and total energy of a photon.

The photon has no mass and always travels at the light speed. From the viewpoint of Newtonian physics, an object without mass does not have momentum and kinetic energy because

$$p = mu$$

$$K = \frac{1}{2}mu^2$$

In the Relativity, things are different as we can tell from Equation (5.4.6) that

$$E^2 = (pc)^2$$

$$E = pc$$

Since the photon has no mass, its energy is the kinetic energy; therefore,

$$K = pc$$

Thus, the kinetic energy and momentum of a photon can be determined from the measured energy.

A particle without mass always travels at the speed of light. For a particle with mass

$$p = \gamma m_0 u$$

$$p = \frac{1}{\sqrt{1 - \left(\frac{u}{c}\right)^2}} m_0 u$$

If u approaches c, the momentum will approaches infinity, and so is the total energy. Therefore, the speed of a particle with mass can never reach the light speed.

5.5 Lorentz Transformations for Momentum and Energy

Let us take a look of Equation (5.4.6) again

$$E^2 = E_0^2 + (pc)^2 \quad (5.4.6)$$

The momentum in general has three components: p_x, p_y, and p_z. Thus,

$$p^2 = p_x^2 + p_y^2 + p_z^2 \quad (5.5.1)$$

Inserting Equation (5.5.1) into Equation (5.4.6) and reorganizing we obtain

$$\left(\frac{E_0}{c}\right)^2 = \left(\frac{E}{c}\right)^2 - (p_x^2 + p_y^2 + p_z^2) \quad (5.5.2)$$

Since the left term is an invariant, i.e. having the same value in all inertial systems; the right side must be an invariant too. Thus, in a system S',

$$\left(\frac{E_0}{c}\right)^2 = \left(\frac{E'}{c}\right)^2 - (p'_x{}^2 + p'_y{}^2 + p'_z{}^2) \quad (5.5.3)$$

The Lorentz transformations for energy and momentum can be derived from the Lorentz transformations for velocity. The equations for transforming from S to S', which is moving at a speed v in the x-direction with respect to S, are

$$p'_x = \frac{p_x - v\left(\frac{E}{c^2}\right)}{\sqrt{1 - \left(\frac{v}{c}\right)^2}} \qquad (5.5.4)$$

$$p'_y = p_y \qquad (5.5.5)$$

$$p'_z = p_z \qquad (5.5.6)$$

$$E' = \frac{E - vp_x}{\sqrt{1 - \left(\frac{v}{c}\right)^2}} \qquad (5.5.7)$$

If momentum is replaced with space location and energy is replaced with time, Equations (5.5.4) - (5.5.7) become Equations (3.1.16) - (3.1.19). This should not be too surprising since energy is related to time and momentum is related to space. Time is frozen if the energy of everything in the universe does not change. An object cannot change space location if its momentum is zero.

5.6 Form of Newton's Second Law

Newton's second law states that

$$F = ma \qquad (5.6.1)$$

In Newtonian physics, mass is a constant and is equal to the rest mass, and Equation (5.6.1) can be written as

$$F = m_0 a \qquad (5.6.2)$$

To examine the validity of the above expression in the Relativity, we start from the general definition of force, i.e.

$$F = \frac{dp}{dt} \quad (5.6.3)$$

Using the definition of p in the Relativity we obtain

$$F = \frac{d}{dt}(\gamma m_0 u)$$

$$F = m_0 u \frac{d\gamma}{dt} + m_0 \gamma \frac{du}{dt}$$

$$F = \gamma m_0 a + m_0 u \frac{d\gamma}{dt} \quad (5.6.4)$$

Equation (5.6.4) is very different from Equations (5.6.1) and (5.6.2). At low speeds, Equation (5.6.4) reduces to Equation (5.6.2); but at very high speeds, the value of γ can be large and the conventional form of Newton's second law is not valid.

Chapter 6: The Twin Paradox

A consequence of Lorentz transformations is that moving clocks run slow. But for two clocks moving with respect to each other, either one of them can be regarded as stationary and the other one as moving. For example, to observers on a train, the clocks on the platform of a train station are moving; whereas to observers on the platform, the clocks on the train are moving. Do we have the answer if one asks "which clocks run slower, those on the train or those on the platform?" A famous debate on this subject is the so-called "twin paradox".

6.1 Space-time Diagram and World Line

A space-time diagram is a coordinate system with time and space as the axes. The diagram has four axes: one for time and three for space. The locus of motion of an object on a space-time diagram is called the world line of the object. The world line of an object can be used to find out the proper time between two events involving the object in a specific inertial system.

Shown in Fig. 6.1 is a space-time diagram with several world lines on it. The space-time diagram here has only one axis for space as motion of all objects is limited to the x-direction. The world lines on the diagram represent:

A: An object stays at a point on the x-axis.

B: An object moves in the x-direction with constant speed.

C: An object moves in the x-direction with varying speed.

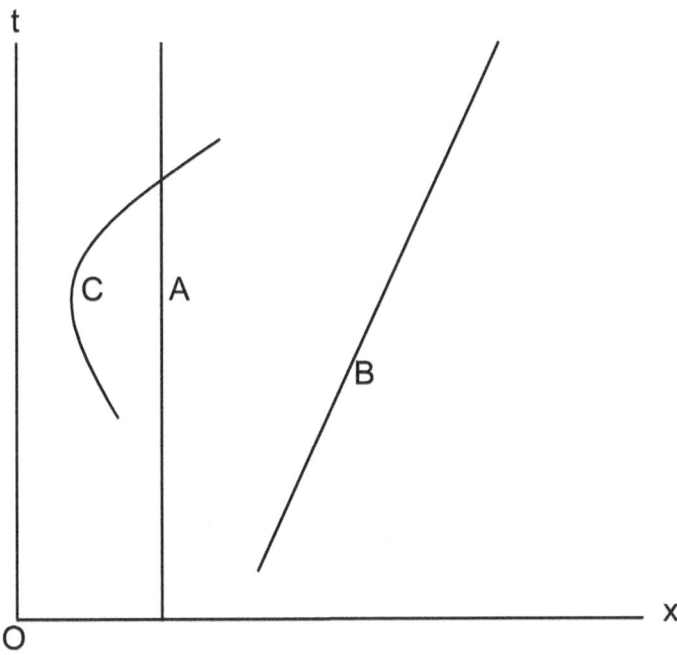

Fig. 6.1 Space-time diagram and world lines

6.2 The Twin Paradox

The paradox is about aging of a pair of twins, brother and sister. The brother always stayed on the earth. The sister, right after her birth, was on board a spaceship traveling along a straight line at a speed of 0.8c. On the 50th birthday of the brother, the spaceship turned and flied back. The world lines of the twins are shown in Fig. 6.2 in which the x-axis is chosen to be coincident with the travel direction of the spaceship. Line OB is the world line of the brother and OAB is that of the sister. At the time of reunion, the brother is 100 years old. How old is the sister?

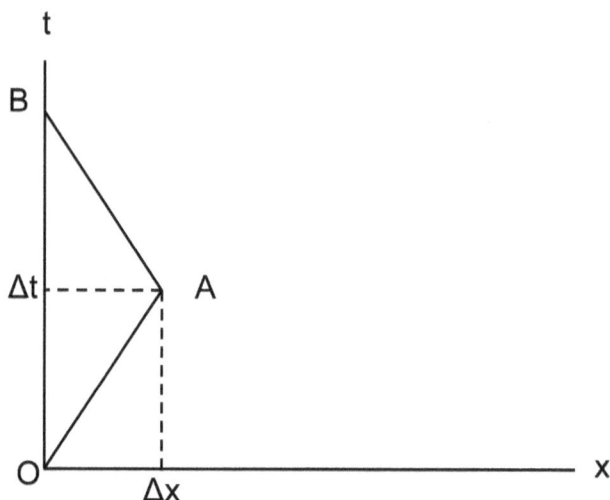

Fig. 6.2 World lines of the twins

Theory of Special Relativity for Beginners

As far as aging is concerned, human bodies behave like biological clocks. As moving clocks run slower, people in motion should age slower. Since the sister was moving all the time, she should age slower than her twin brother. However, it can be argued that the earth was moving with respect to the spaceship, and that the brother instead should age slower. But this argument was opposed by the point that prediction of physical phenomena must be done in an inertial system, and that the spaceship is not an inertial system as it not only changed direction once but also decelerated and accelerated during the course of changing direction. Thus, the only valid inertial system available for making the prediction is the earth system.

The sister's journey can be divided into two parts: the first half and the second half. If we ignore the small effect of deceleration and acceleration during the course of changing direction, then in each part the spaceship system is an inertial system. If two events correspond to the beginning and end of the first half of the journey; we can find the proper time of the two events, and the proper time is equal to the time duration read from a clock on the spaceship. Although we don't know the time and space intervals between the two events in the spaceship system, we do know the time and space intervals in the earth system. Thus, we can use the time and space intervals in the earth system to calculate the proper time of the two events. Since the proper time corresponds to the time duration read from a clock on the spaceship, it also corresponds to the sister's age.

The proper time $\Delta\tau$ between the two events is

$$\Delta\tau = \sqrt{(\Delta t)^2 - \left(\frac{\Delta x}{c}\right)^2}$$

where Δx and Δt are space and time intervals obtained in the earth system. Since the spaceship traveled at a speed of 0.8c,

$$\Delta x = 0.8c\Delta t$$

Thus,

$$\Delta\tau = 0.6\Delta t$$

We know that Δt correspond to the age of the brother. Thus, the age of the sister is only 60% of that of the brother. The situation of the second half is the same as the first half. We can conclude that the brother is 100 years old and the sister is 60 years old at the time of reunion.

Is the paradox really resolved? If the spaceship did not return and kept flying in the same direction, then the spaceship is moving with a uniform velocity with respect to the earth, and the spaceship system is also a valid inertial system. Now it can be argued that the earth is moving and the brother instead should age slower. Some physicists then argued that the spaceship is the very one that is moving since it had to accelerate in order to reach the speed of 0.8c; therefore, the sister should age slower. This argument suggests that clocks with higher absolute speed run slower than those with lower absolute speed, and has been supported by experiments.

As it is the absolute speed that determines the time rate change of a clock, then what is the essence of time dilation resulted from the Lorentz transformations?

6.3 Essence of Time Dilation and Space Contraction

It was said in Chapter 3 that "moving clocks run slow" is a consequence of the Lorentz transformations. But it was pointed out in the last section that clocks with higher absolute speed run slower. Although we cannot determine the absolute speed of an object, we can determine if an object has higher absolute speed and by how much than other objects. For example, we know that the sister, who traveled with the spaceship, has higher absolute speed than her twin brother; therefore, the sister aged slower than her twin brother.

The Lorentz transformations deal with the relative speed of two inertial systems, not with the absolute speed of an object. Thus, "moving clocks run slow" is not attributed to time dilation. Similarly, "moving rods contract" is not attributed to space contraction. In fact, moving rods do not contract. The contraction theory proposed by George Francis Fitzgerald was rejected because there is no experimental evidence of length contraction at all.

Time dilation and space contraction is a measurement effect and is due to the relative motion between inertial systems. Are they real? Yes, they are. However, we should keep in mind that they have nothing to do with real clock rate change and real length change.

References

Basic Concepts in Relativity; Robert Resnick and David Halliday, (MacMillan Publishing Company, New York, 1992)

Elementary Modern Physics, Third Edition; Richard T. Weldner and Robert L. Sells, (Allyn and Bacon, Inc., Boston, 1980)

Einstein's Theory of Relativity; Max Born, (Dover Publications, Inc., New York, 1962)

Relativity for the Million; Martin Gardner, (The Macmillan Company, New York, 1962)

Appendix A: Another Book by the Author

Four-Element Model

of

Particle Physics

Everything in our universe is made from four basic elements: negative charge, positive charge, n-bit, and p-bit.

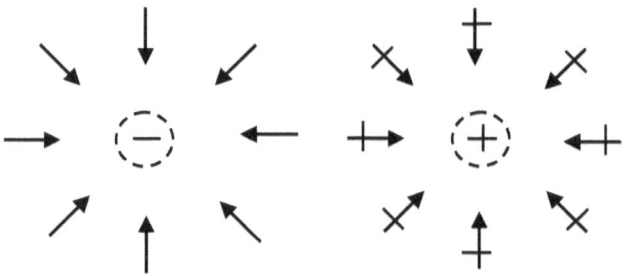

Jen Chyi Liao

Ancient Greek philosophers introduced a ''Four Element Theory'' saying that the four most basic elements of our universe are earth, water, air and fire. From modern view of particle physics, the most basic elements of our universe are the so-called fundamental particles such as electrons and quarks. In this book, the author presents a ''Four-element Model'' with a novel idea that the most basic elements of our universe are: positive charge, negative charge, and two tiny energy carriers named p-bit and n-bit.

The underlying concept of the four-element model is that everything in our universe is made from the above-mentioned four basic elements. All forces between particles are the result of interactions among these elements. Mass and gravity have the same origin based on this concept.

The four-element model consists of the following postulates:

Postulate1

The four elements of the universe are: positive charge, negative charge, n-bit and p-bit. The n-bit and p-bit are energy carriers of the universe. The n-bit interacts only with the negative charge, and the p-bit interacts only with the positive charge. The negative charge does not interact with the positive charge. The n-bit does not interact with the p-bit.

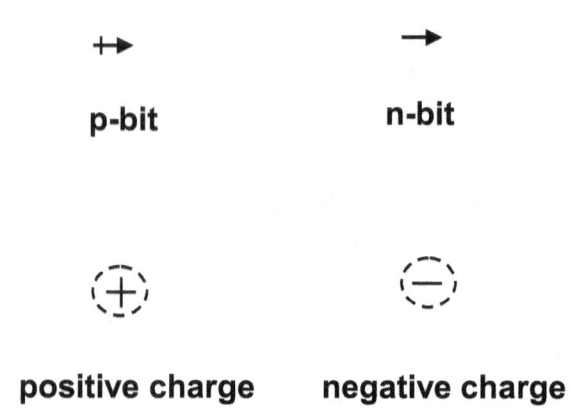

Fig. A1 The four elements

Postulate 2

An electron-neutrino is formed by a group of n-bits, and an anti-electron-neutrino is formed by a group of p-bits.

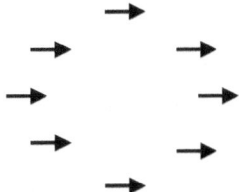

Electron-neutrino

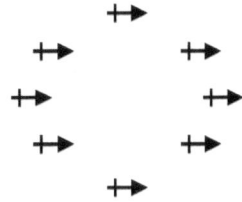

Anti-electron-neutrino

Fig. A2 Electron-neutrino and anti-electron-neutrino

Postulate 3

A photon is a composite of an electron-neutrino and an anti-electron-neutrino.

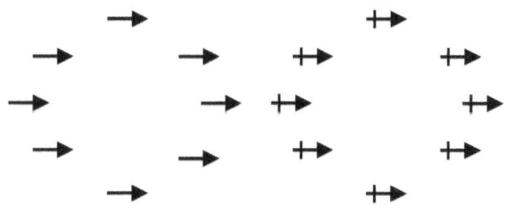

Fig. A3 Photon

Postulate 4

An electron contains a negative charge and an electron-neutrino. A positron contains a positive charge and an anti-electron-neutrino.

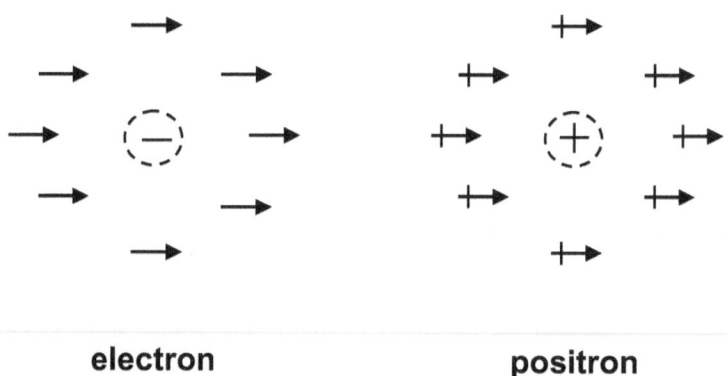

electron   positron

Fig. A4 Electron and positron

Postulate 5

There are bits and charge pairs in the vacuum. A charge pair consists of a positive charge and a negative charge. The charge pair is electric neutral because the two charges are close to each other.

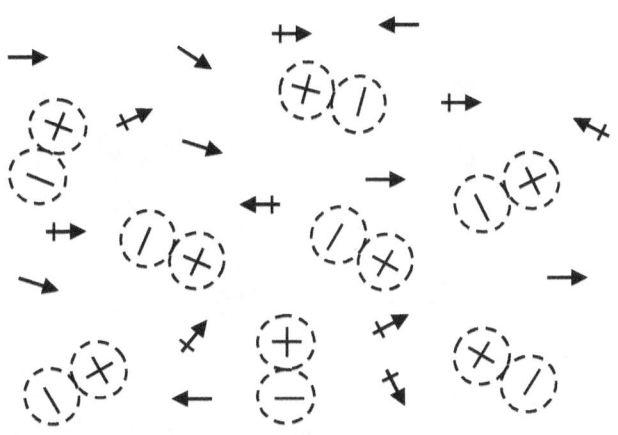

Fig. A5 Bits and charge pairs

Postulate 6

Mass is a joint effect of charges and bits.

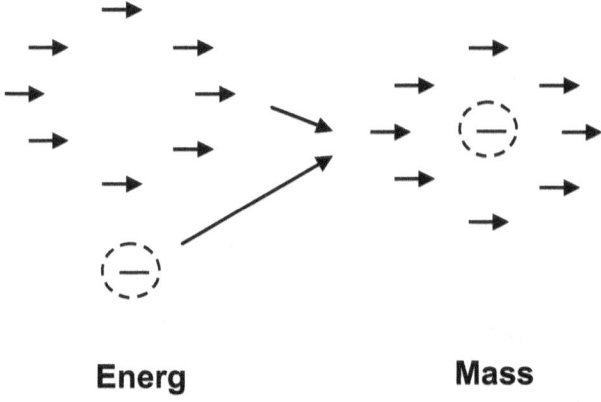

Energ   **Mass**

Fig. A6 Energy to mass

Postulate 7

All forces are the results of interactions among the four fundamental elements.

Postulate 8

The n-bit is the anti-element of the p-bit, and vice versa. The negative charge is the anti-element of the positive charge, and vice versa. Replacing every constituent element with its anti-element changes a particle into its antiparticle.

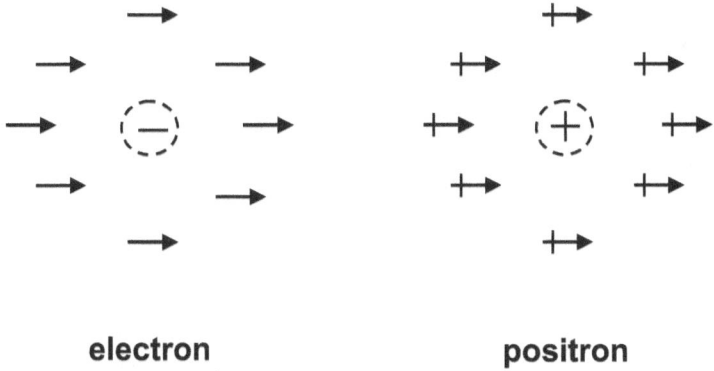

electron   **positron**

Fig. A7 Anti-element and antiparticle